THE APPLE BOOK

T·H·E
APPLE
B·O·O·K

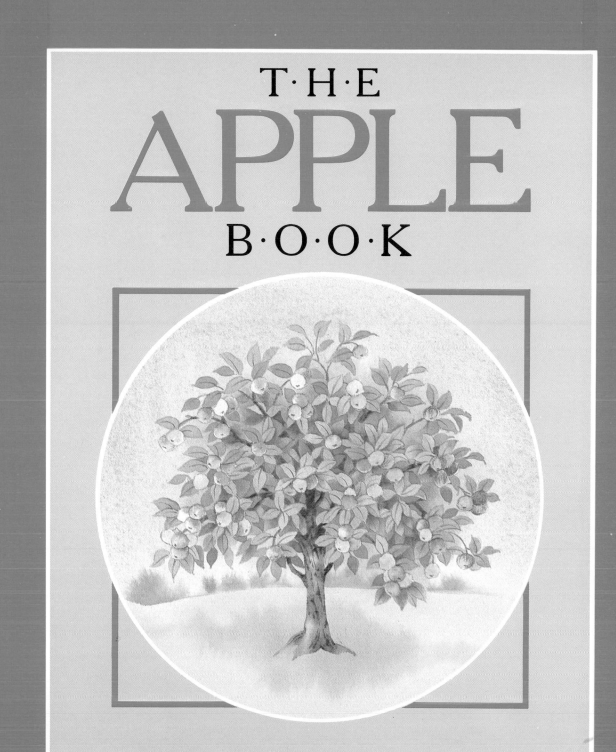

PETER BLACKBURNE – MAZE NDH

COLLINGRIDGE

Front cover illustration by Jenny Mitchell
Back cover photograph Lindley Library/Eileen Tweedy
Recipes contributed by Bridget Jones
Illustrations by Cynthia Pow

The author would like to extend his especial thanks to
Hugh Ermen of The National Fruit Trials, Brogdale
Experimental Horticulture Station, to Brian Self of East Malling
Research Station, and to Harry Baker of the Royal Horticultural
Society, Wisley; all of whom haved helped him with photographs
and information.
He would also like to thank all those who have taught him about
fruit growing, especially A.G.H.

Published in 1986 by Collingridge Books
an imprint of The Hamlyn Publishing Group Limited,
Bridge House, 69 London Road, Twickenham, Middlesex, England

ISBN 0 600 30689 5

Filmset in England by Vision Typesetting, Manchester
in 10 on 11 pt Garamond

Printed in Spain

Contents

Introduction

ne of our best-loved fruits, the apple has been with us for as long as we can remember. It recurs throughout history, playing its part in legend and folk tales, appearing in Greek mythology as well as in the Bible. Famous names have been associated with it, and apart from its undisputed value as a food and the basis of many drinks, it has come to stand as a symbol of health and purity.

The precise origin of our modern apple remains obscure, but this is not unusual in the botanical world, where many plants have the same blurred beginnings. But we know that it has been in cultivation for a very long time, and indeed there is archaeological evidence of its use as a food from as long ago as six centuries BC.

The story of the development of its cultivation in Europe is a fascinating one, taking us back to Roman days when the advancing armies of the Roman Empire carried with them fruits and seeds to plant in the lands they conquered. As knowledge spread and better use was made of natural resources, so the secrets of fertilization and cross-breeding began to unfold and by the time of the Norman Conquest in Britain different varieties of apple were already acquiring identifiable characteristics. How theory and practice, trial and error, continued through the centuries to bring us the variety of fruits we know today provides an absorbing record of painstaking research.

Step by step, improvements have been made in the selection of rootstocks, methods of pruning and training, and choice of tree forms. Sometimes there have been setbacks, sometimes a lucky breakthrough when a chance seedling proves to be a classic.

The possibilities have never been more promising than they are today. New techniques, unheard of a few years ago, using hormones and controlled radiation, and experiments in tissue culture, are leading to improved stocks and virus-free strains. There are varieties now available to suit all tastes, and modern facilities for marketing and transport ensure that there is a wide choice throughout the year.

The future looks exciting – perhaps the time is not too far off when we may achieve the goal that has always been thought impossible, and finally produce the ideal apple.

I

Myths, Legends and Facts

pples, it is fair to assume, have been an important part of our diet ever since the development of early man; they were probably consumed by his ancestors, too. In the period of prehistory it was just a matter of gathering the fruit and eating it when you felt hungry – the original hand-to-mouth existence! Small wonder that the apple has played such a prominent part in folklore and legend, featuring in Greek myths and representing Temptation in the Bible story of Adam and Eve.

One of the hardest things when writing any kind of history is to establish whether what you are writing is fact, fiction or a mixture of the two. Take Sir Isaac Newton, for example. That he existed is not in dispute, but how much credence are we to give to the story that the apple falling in his garden was the inspiration behind his declaration of the laws of gravity? It makes a splendid story, so it is well worth relating, but is it true?

Then again, were William Tell, the Swiss national hero, and Johnny Appleseed, of American folklore, real people or fictitious characters? It may surprise you to know that the story of William Tell is pure fiction but Johnny Appleseed most certainly lived. Another question is the true identity of the 'apple' that so often crops up in these old and translated stories. Was it really an apple or some strange foreign fruit that readers would never have heard of?

Opposite *The Garden of the Hesperides*

The Garden of the Hesperides

Of all the ancient myths, some of the best known are those which describe the twelve labours of Hercules. The particular labour that interests us is his eleventh concerning the garden of the Hesperides. This had been given to Juno as a wedding present by Jupiter when they were married (or to Hera by Zeus if you are a Greek scholar).

This unusual garden was looked after by three nymphs, the Hesperides, and in the middle of it was an apple tree which bore golden fruits. The tree was guarded by a great dragon that lay awake the whole time with his tail wound round the trunk. Presumably he had a pretty foul temper as well because woe betide anyone who had ideas about stealing those apples.

To find the garden, Hercules consulted a sea-god called Nereus, a creature who could turn himself into anything he wished. Hercules extracted the information from him without too much difficulty and after a long and difficult journey, he found the garden, guarded by its nymphs and dragon.

At that point the story rather loses direction because the Hesperides, instead of turning Hercules to stone, simply allowed him to take a couple of the fruits; nor did the dragon seem to raise any objection. It ended with the apples starting to tarnish and being taken back to the garden where they regained their former lustre.

An interesting horticultural point concerning the identity of the tree keeps turning up in mythology. There is a general feeling that 'apple' is simply the word that a lot of translators used when faced with a fruit that was either unfamiliar to them or likely to be so to their readers. Although apples grew in the Mediterranean region in those times, the pomegranate was far more common and apart from this, its yellowish-orange colour could easily be interpreted or mistaken for gold. It is equally possible that 'golden apples' were just oranges or lemons.

Atalanta's race

They feature, too, in the story of Atalanta, who was notoriously fleet of foot. To select a suitor, she announced that she would marry anyone who could beat her in a race, but if he failed to win he would pay for it with his life. One by one her suitors tried and lost, forfeiting their lives. Finally one young man, Hippomenes, who was determined to survive and win Atlanta's hand, appealed to Venus, the goddess of love. She gave him three golden apples and with these he entered the race. As Atlanta gained on him, he threw one of the apples a little ahead of her, causing her to pause long enough for him to pass her. Three times this happened, and the third time he ran ahead to win the race and so claim his bride.

The 'golden' theme appears everywhere. Wasn't it a golden apple that was supposed to have started the Trojan war? Someone called Eris threw one into the middle of a reception out of pique because he

OPPOSITE Adam and Eve in the Garden of Eden *by Lucas Cranach (1472–1553)*

Atalanta and Hippomenes by Guido Reni (1575–1642)

hadn't been invited to the wedding (see page 19). That particular golden apple had written on it 'For the fairest', and this is reminiscent of a charming ancient Greek custom. If a young man was rather shy about proposing to a girl he could throw an apple to her as an alternative to asking the question. If the girl wanted to accept, she would then catch the apple.

Wassailing

Apples are associated with fertility rites in the odd but widespread custom of wassailing. This goes back into the mists of time and may even date from the Romans as part of their festival for praising and encouraging Pomona, the goddess of fruit trees. What is certain is that it was originally carried out on Twelfth Night to ensure a good crop of apples in the coming year. The practice seems to have been extended beyond Twelfth Night and apples because it came to mean more or less anything that was done in the name of fertility.

The original wassailing was popular in the great English cider counties of Gloucestershire, Herefordshire and Somerset and is still carried out in some villages. There is a definite ritual and form to the festivities. The largest tree in the orchard, or village, is chosen for the site of the party and then pieces of toast are hung in the branches. This is done to attract robins to the tree as these are held to be the

good spirits of the orchard. It not only attracts them, it also feeds them in the depths of winter. To ward off the evil spirits, the villagers stand round the orchard and blast off guns into the branches.

Then cider is brought out. Some of it is poured on the tree roots but most of it seems to disappear down the throats of the participants. Accompanying this thinly disguised orgy is a dance that is carried out round the tree. No one really seems to know why they are supposed to be doing it but the result is the same – a good time is had by everyone.

Apple trees being toasted with hot cider in Devonshire on the eve of Twelfth Night (1861)

The legend of
William Tell

Of all the tales of people connected with the apple, one of the most familiar is that of the Swiss folk hero, William Tell. The story is so well known that most people regard it as history. In fact it is pure mythology, in much the same way that Robin Hood is in England. Like Robin Hood, the story is probably based on a mixture of tales woven around half-truths.

Not only was William Tell reputed to be one of the best crossbowmen in Switzerland, he was also recognized as the finest oarsman on Lake Uri. At that time Switzerland was under the rule of Austria and there was a lot of unrest in the country with much antagonism against the ruling power. William Tell, good patriot that he was, resisted whenever he could.

Unfortunately for him, his district was overseen by a tyrant called Gessler who had decreed that anyone who did not salute the Austrian standards should die. William Tell had always steadfastly refused to pay homage to any foreign emblem and when Gessler caught him openly flouting his orders, Tell was immediately arrested.

In front of all the inhabitants of the town, his son was fetched and tied to a tree and, to their horror, an apple was put on the boy's head. The implications were clear. William Tell took his crossbow and two bolts; one he fitted into the bow, the other he put in his belt. The first bolt split the apple cleanly in two and buried itself in the tree. With customary bravado, William Tell then turned to Gessler and said, 'If I'd hurt the boy, the other bolt would have been in your heart.'

True to form, Gessler had left his options open and instead of death, therefore, William was condemned to prison. While they were sailing across the lake to the prison a storm sprang up that threatened

William Tell shooting at the apple placed on the head of his son. From Henry Petri's edition of Munster's Cosmography *(1554)*

to capsize the boat. Known as a skilful sailor, William Tell was immediately untied and soon had the boat back under control. As he neared the far bank, he purposely took the boat a little closer than usual to some rocks and, at just the right moment, he sprang ashore.

When word of the escape reached Gessler, he threatened to kill the whole Tell family if William didn't give himself up at once. But William set an ambush for Gessler on the road to the prison and thus put an end to the tyrant.

The story goes that the Swiss were so inspired by this act of heroism that they rose against the Austrians and didn't stop fighting until the last one had been driven out of the country. William Tell was then offered the crown of Switzerland, but, like all true heroes, he gratefully turned down the offer and went back to his beloved mountains where he lived happily ever after.

A descendant of Sir Isaac Newton's original apple tree, which grows today in the Cambridge University Botanic Garden

Sir Isaac Newton and the laws of gravity

There is nothing legendary about Sir Isaac Newton (1642–1727), who was a renowned mathematician and a thinker. His discovery and application of the laws of gravity were of incalculable value to astronomy. Tradition has it that these laws were the end product of a train of thought which started when he saw an apple fall from a tree in his garden at Woolsthorpe Manor between Grantham and Belvoir.

Isaac Newton watching an apple fall at Woolsthorpe Manor. This incident was reputedly the inspiration behind Newton's laws of gravity

Not surprisingly, the tree that is thought to be the one in question no longer exists but direct descendants of it are still to be found. One is in Cambridge University Botanic Gardens and another in the National Fruit Trials apple collection at Brogdale Experimental Horticulture Station near Faversham in Kent. A notice beneath the Cambridge tree announces that 'This apple tree is a descendent by vegetative propagation of a tree which grew in the garden of Woolsthorpe Manor, near Grantham, and which is reputed to be the tree from which fell the apple that helped Newton to formulate his theory of gravitation. The original tree is said to have died about 1815–1820. The variety is Flower of Kent.'

Besides the spelling mistake, the most interesting thing about this notice is the variety of the tree, which is by no means certain. When identification was undertaken, there was no fruit sample or accurate picture of Flower of Kent with which to compare the apple. It was simply named on the basis of its appearance, which tallied with the description of Flower of Kent. Quite rightly, it was later declared to be a variety in its own right and named 'Isaac Newton's Tree'.

Johnny Appleseed

The basis of the story of Johnny Appleseed that we know today is perfectly true; only the details have changed a little with the passage of time. His real name was John Chapman and he was born in Leominster, Massachusetts, U.S.A., on September 26th, 1774. The

Johnny Appleseed scattering his apple seeds across America

story goes that he wandered the United States far and wide with a satchel slung over his shoulder, scattering apple seeds wherever the spirit willed him. That is romantic licence which does him less than credit. He would certainly have known that propagating apples from seed was a futile pastime because they would never have come true to type.

What he actually did was to establish a string of apple nurseries from Pennsylvania in the east, through Ohio and into Indiana in the west. This he did very successfully between about 1800 and the year of his death in 1845. By the end of his life, New England had a thriving fruit export business, even sending apples to the West Indies.

Another myth that has to be exploded is that John Chapman was the father of modern apple growing in America. Certainly he was a colourful character and did an enormous amount to spread the growing of apples in the eastern states, but the centre of the apple industry in America is in the north-west state of Washington, some 1,600 miles west of Iowa. Nor can it be said that the first commercial apple was one of Chapman's. It was very probably one that had come direct from England, as had all those original ones in New England.

*Red Delicious, an offspring of
Jesse Hiatt's Hawkeye apple*

▨ Other pioneers in ▨ America

It is believed that one of the first cultivated apples to be grown in Washington State was sown by a Captain Simpson at Fort Vancouver in 1824. This was grown from the seed of an apple that the Captain had eaten at his farewell dinner party in England. It was the only seed of the batch that came to anything and was probably the tree which initiated the setting up of the commercial orchards. It is a pity that no one took the trouble to remember the variety.

The two men most responsible for the growth of the apple industry in the United States were Henderson Luelling and William Meek, both from Iowa, though at first they did not know of each other's existence. Along with other fortune seekers, Luelling set forth in his covered waggon to 'go west'; he, full of hope, the waggon full of soil and apple trees. With his waggon a good deal heavier and more cumbersome than the others he was soon left behind, but luckily for him the Indians left him alone. He reached Washington State where he met with Meek and, together, they set about planting orchards.

THE · JUDGEMENT · OF · PARIS

In Greek mythology the story goes that Eris, God of Discord, was not invited to the nuptial feast of Thetis and Peleus and, in his chagrin, he threw a golden apple into the wedding gathering 'for the fairest'. The apple was claimed by Hera, Athene and Aphrodite and in order that the matter be settled, Zeus decided the three goddesses should put their arguments to a mortal man; Paris, son of King Priam of Troy, was chosen. In turn the goddesses tried to persuade Paris, offering him power and victory in batle and, from Aphrodite, goddess of love, the most beautiful woman in the world. Paris chose to award the golden apple to Aphrodite and was rewarded with Helen, wife of Menelaus King of the Greeks.

The anger of Hera and Athene was united against the family of Paris and so began the long and bloody Trojan wars which ended in the eventual destruction of Troy.

As it happened, they picked a particularly good time to be setting up a food business, for those were the days of the Gold Rush when every town was full of people trying to strike it rich. By the time the demand for apples in the western states began to decline, the railway had stretched across the Rockies and this made it possible for the fruit to reach anywhere in the whole vast continent. Washington State became the largest apple-producing area in the world.

Something was also taking place in Iowa that was to sweep the apple-growing world. In a field near the town of Peru, Madison County, there stands an inscribed granite memorial commemorating an apple, the only such memorial known to exist. In the mid-1800s a Quaker farmer, Jesse Hiatt, noticed that a sucker had sprung up from the rootstock of a tree that had died. This grew into a tree but the apples that it carried were of a completely different kind – medium sized, bright red and with excellent flavour and aroma.

Hiatt named the apple Hawkeye and for twenty years or so the tree grew and prospered. However, when he sent the apples to a fruit show in 1893, the judge, on biting into one, exclaimed 'Delicious, delicious!' In 1895 it was introduced to the trade as Delicious by Stark Bros. and there started a career which was to lead to Delicious being the most widely grown apple in the world. By the mid-1920s its offspring were said to number between seven and eight million. Several sports or variants have sprung from it, the most famous being Red Delicious, which has easily replaced it for popularity in America.

Oddly enough, Golden Delicious has no connection with Delicious at all. This was found as a chance seedling in West Virginia in about 1890 and was introduced, again by Starks, in 1914. It was probably given its name with the hindsight of what had happened to the original Delicious. Unfortunately, as is sometimes the case when propagation is carried on to an unlimited extent, changes have taken place in the apple so that today it is pretty certain that we do not have an exact likeness of the original.

OPPOSITE *Apple picking in Baltimore, America, between the wars*

The History of the Apple

The modern cultivated apple has no naturally occurring counterpart in the wild, as it has been made up of selections and mixtures of wild crab apples over the centuries. It is generally accepted that the apple as we know it is a descendant of the wild *Malus pumila*, the original *Pyrus Malus* as named by Linnaeus. This is a crab apple that is found growing wild in Europe (including Britain) and western Asia as far east as the foothills of the Himalayas.

Malus pumila is extremely variable in the wild with fruit ranging in colour from yellow to bright red and with a habit of growth varying from almost as upright as a Lombardy poplar to weeping. Probably the best selection from the original is the cultivated crab 'John Downie'.

However, by no stretch of the imagination could one relate a crab apple, even as variable as *Malus pumila*, to something with a fruit as large as a Bramley's Seedling or as tasty as a Cox. Clearly, there are other fingers in the apple pie and it is felt that the most likely intruders are probably *M. prunifolia* and *M. silvestris*. Both these species are quite commonly found growing wild throughout the temperate zones and *M. silvestris* is known as the wild crab apple in Britain.

From this, you will begin to see why the exact origin of our modern apple is so obscure, and as well as the natural cross-breeding that goes on in the wild, mankind has interfered in the selection process as well. For, not content with the natural confusion that exists, we have ourselves introduced yet another species into the picture, the Siberian Crab, *Malus baccata*. As the name implies, this species comes from the Siberian region of U.S.S.R. so it is particularly well suited to withstand cold conditions. It is for this quality and the virtue of its being much less susceptible than other crabs to the fungus disease apple scab, that it has been used in breeding hardier and healthier apples.

So the apple's family tree is extremely complex and more or less any theory that a botanist cares to put forward is difficult to disprove. However, this account is the one most widely accepted today on the present evidence.

The apple as food

Though for thousands of years the fruit must have been picked and eaten off the tree, the first definite evidence of apples having been collected and brought home as food was discovered in what is now part of Turkey, where the remains of fruits have been found dating back to around 6,500 BC. These fruits would almost certainly have been gathered and not picked from cultivated trees.

In Britain apples certainly seem to have been used since Neolithic times (the New Stone Age) as evidence of pips have turned up on several sites. In Switzerland, excavations of prehistoric dwellings have yielded the remains of what are thought to be dried half-apples. This suggests that the inhabitants had a knowledge of how to store the fruits for out-of-season use. This method of storage also seems to have been quite widespread in the regions adjoining the Caucasus mountains, where wild crab apples were plentiful.

This, in fact, is one of the driving forces behind the cultivation and subsequent treatment of all food plants. First, they have to be selected, improved upon and then grown under better-than-natural conditions. Once that has been achieved, thoughts can turn to ways in which the crop may be preserved for later use.

This has two obvious benefits. The first is that the crops can be grown in quantities far in excess of immediate needs, in the knowledge that they will not be wasted. Secondly, it allows the surplus to be stored beyond the season when it is available fresh, thus ensuring that the supply of food is maintained.

This may sound very elementary to us in our world of freezers and controlled atmosphere stores but something even as simple as dried apple rings was a real breakthrough in a world where fruit tended to last no longer than the landing of the first fungus spore upon it.

The selection and development of improved varieties is thought to have started with civilizations such as the Greeks and Romans, though we cannot be certain of this.

The spread of the apple

It was mainly the Middle Eastern trade routes and the sort of prehistoric Common Market which then existed that were responsible for the spread of the apple from its original home. Merchants and travellers were journeying far afield, and, with them, they took either seeds or actual plants of their main economic crops — amongst them, apples. In this way, they reached the eastern Mediterranean in about 2,000 BC.

PATCH BUDDING
Cut out and remove a square of bark containing a bud (1 and 2) Remove a flap of bark from the stock (3) Position the 'patch' on the stock (4) and tie in place with raffia

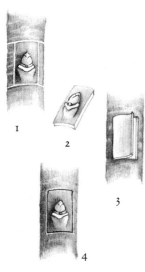

From there, they were taken to Egypt where it is known that apples were in cultivation in the Nile delta during the 12th and 13th centuries BC. They not only spread south to Egypt, however, but also west to Greece and Italy. Several references exist which prove this, the earliest being in Homer's *Odyssey* (900–800 BC).

At that time the cost of apple fruits was prohibitive which is only to be expected with something so new. The rarity value of apples was certainly linked to the scarcity of trees, something that was primarily brought on by the uncertain methods of propagation that existed in those days. The earliest orchards would have consisted of trees raised from pips and, provided that the farmers or gardeners were content with the original species, the seedlings would have continued to come true to type, developing into trees near enough the same as their parents. It was when new selections and hybrids came to be used that the difficulties started. Anyone who has grown an apple tree from a pip will have noticed the same thing that these ancient plant breeders found: that the progeny were not the same as the parent. This was an early lesson in the science of genetics.

In practical and modern terms it means that if trees are raised from the seeds of recent hybrids, they will not come true to type. The laws of Mendel can, to some extent, predict just what turns up but you may get a lot of throwbacks from previous generations. If, therefore, the intention is to propagate a named variety of apple so that you end up with another tree exactly the same, the only way to do it is by vegetative methods. That is as true today as it was all those years ago.

With most plants, the simplest form of vegetative propagation is by taking cuttings. Not so with apples. Unfortunately, apple cuttings, even of the same variety, are very erratic when attempts are made to strike them by what might be called 'traditional' methods. Some will form roots readily, others will take much longer and many will not root at all but simply rot off. Those early growers knew this to their cost but, out of necessity, they found a way round the problem.

Like many other discoveries, it could well have been keen observation that gave the clue rather than years of painstaking work. In nature, one frequently sees examples of two branches having joined themselves together, or become fused, simply by being pressed against each other for what may have been several years. It happens quite commonly in beech hedges. Now, if this phenomenon can occur all by itself in the wild, surely it can be made use of in cultivation. It can indeed, and it is known as the practice of grafting.

❧ *Grafting* ❧

Grafting and budding (a form of grafting but on a much smaller scale) are still the main ways of raising fruit trees vegetatively. Some of the materials used now are more sophisticated than they were then but the operation is exactly the same. This is hardly surprising; after all, the object of grafting hasn't changed. What is wanted is a firm and lasting union between a woody shoot of a selected variety (the scion) and a ready-made set of roots (the rootstock).

Illustration of tools and budding and grafting techniques from Paradisi in Sole *by John Parkinson (1629)*

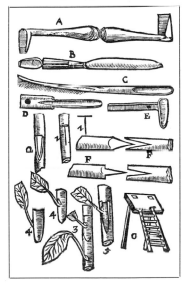

Bearing in mind that we want the two parts to grow together, identically sized and shaped cuts are made on the scion and the rootstock so that when they are brought together they fit exactly.

This 'union' is then bound with either raffia or a special plastic tape and then waxed over. The raffia holds the union firm and the wax prevents it drying out. The graft is best carried out in spring when the sap is starting to flow and, by midsummer, the union is normally firm enough to allow the binding to be removed.

That is really all there is to basic grafting. However, a refinement was perfected at about the same time, and that was 'budding'.

Budding

A form of patch budding from The Booke of Arte and Maner, Howe to Plant and Graffe all Sortes of Trees *by Leonard Mascall (1572)*

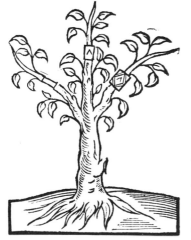

Whereas grafting involves the use of a short length of scion shoot containing two to three buds, in budding only a single bud is used. Also, instead of the bud being part of a shoot, it is entirely removed from the parent plant and slid into a waiting slit in the bark of the rootstock some 15 cm (6 in) from the ground. To be absolutely correct, the method described here is a slight modification of the original but is the one that is widely used now.

In those far-off times, the method was to remove a complete square or circle of bark from the scion variety. This would have a strong bud in the centre of it and the whole thing was placed on an identical-sized 'wound' on the rootstock.

Anyone concerned with the nursery industry today will recognize this as a very accurate description of *patch budding*, a system which has recently been rediscovered and is being used increasingly widely. Nothing changes, even if it does take the odd two thousand years to come round again!

ABOVE *Roman wall painting, possibly depicting apple trees, from the villa of the Empress Livia at Prima Porta, Roma* LEFT *Fruit picking in Roman France. A mosaic from St Roman en Laye*

Glastonbury Tor or Avalon – a place where apples grew

◦ R o m a n   p l a n t   b r e e d e r s ◦

By Greek and Roman times the apple was firmly established in the lands bordering the north coast of the Mediterranean Sea. There it was being propagated successfully by vegetative means (grafting) to produce any number of identical trees.

This was the turning point in apple production because it made it possible for early nurserymen and, later, plant breeders to earn a respectable living. There was now nothing to prevent them from introducing as many varieties as they could, knowing that they could be propagated true to type. The chances are that it led initially to an influx of many useless varieties. The situation would have settled down in time and new varieties would have started to appear on the basis of their excellence rather than on their novelty value. This is confirmed by the writings of such authorities as Virgil, Cato and Pliny, the last of whom listed some two dozen varieties of apple that were available. A contemporary of theirs, Columella, actually described and recommended what were clearly distinct varieties such as the Armerian, the Cestine and the Syrian.

So the system of naming seems to have been much the same as it is today, that is, after either the finder, some benefactor or notable person or after the place from which it came. It is not known if these varieties were the result of deliberate breeding, but it is more likely that they were either naturally occurring variations or man-made selections. In any event, these were the forefathers of the modern cultivated apple.

It is interesting to note, though, that these early growers had much the same problems then as we do today where early-ripening varieties are concerned – they tended to be dry and floury. You only have to think of a slightly over-ripe Beauty of Bath or Early Victoria to see that they were up against the same difficulties.

◦ *The apple in Europe* ◦

With the apple well established and in the hands of the Romans, its spread and future were assured. It marched majestically across Europe with the armies, being planted wherever they settled for any length of time, and eventually crossed the Channel.

But the apple was growing in Britain long before the Romans appeared on the scene. As in the case of Asia Minor, fruit growing in Britain was, in its early days, a pretty haphazard affair and took the form of gathering and storing rather than consciously and intentionally planting an orchard and growing fruit. Evidence of apple pips has been found in settlements dating back to as long ago as the New Stone Age. These were obviously from wild fruits that had been gathered and brought back to the settlement.

There are no written records of what was going on then, but it is known that the Celtic word for apple *abhall* and the Cornish *avall* existed at that time. In fact, it is from *avall* that Avalon (Glastonbury) got its name – a place where apples grew. Interestingly enough, the word for apple is similar in many countries which are near or part of

Boys knocking down apples from Commonplaces of Christian Religion *(1563)*

Britain: besides Celtic and Cornish, the Old English is *aeppel*, the German *Apfel*, the Old Norse *epli*, the Irish *abhal* and the Welsh *afal*.

Once the Romans invaded Britain there is much more evidence that apples were in general use, presumably by them, as everything seems to have become better organized. The evidence of this lies in the many excavations that have unearthed – either actual bits of apple or signs of their presence.

Although there is this, what one might call, circumstantial evidence in favour of organized cultivation before the Norman Conquest, it would seem that there was little to be gained from it. There were probably enough wild apple trees to satisfy the modest needs of those times and there is no firm evidence that the cultivated varieties from Europe came in before 1066.

The Norman Conquest

The arrival of William the Conqueror in 1066 heralded a new era in fruit-growing. The French had almost certainly acquired their knowledge of fruit growing from the Romans and, in this respect, were far more advanced than the Britons. The main reason for this ties in very neatly with the spread of Christianity and the establishment of monasteries across Europe.

The setting up of monasteries in Britain led to an important development in fruit growing which might not be immediately apparent, and the subsequent exchange of knowledge between the English and French establishments undoubtedly led to organized apple growing in Britain for the first time. It was in these largely isolated and self-contained communities that most of the groundwork went on and experience in apple cultivation was gained. They were the places where many new varieties were raised and where improved methods of cultivation were developed. In all probability the monks also looked into the extended use of apples. This includes perhaps the greatest gift from the Normans: the excellent apple spirit calvados.

Cider, on the other hand, had probably been with us since prehistoric times. After all, the brown fluid that frequently oozes out from the bottom of a pile of apples is not as far removed from cider as one might imagine. It would only have taken one inquisitive caveman to taste it for a cottage industry to spring up!

Early varieties in Britain

Among the apple varieties grown at the time of the Conquest there were probably a few left over from those introduced by the Romans, such as the dessert apple Decio, but they would have been few and far between. Most of the true Roman varieties are likely to have been unsuitable for the English climate.

Within the illustration:

Ly commence le cinquieme
liure ¶ Rustican parlant
des arbres ¶ Et de la nature

tux des plantes et des cho
ses communes appertenans
a labourage et de chascune

An orchard as depicted in the 15th century French manuscript, Rustica du Cultivement des Terres

The variety Decio, if genuine (and there is no reason to suppose that it is not), would certainly rank as the oldest apple in Britain. It is said to date from the time of Attila (around AD 450) and to have been brought over by the Roman general Etio. Even if the evidence to support the belief is a little scant, there is absolutely nothing to disprove it and, in any case, it's an attractive story based on a strong probability. Decio is still represented in the National Fruit Trials at Brogdale.

A number of French varieties would have been grown and some of them still survive, but the majority were nondescript seedling trees, whose fruit was really only suitable for cider.

An interesting point is that most of the native English apples, the crab apples, were especially good for cider making but not for cooking or dessert. The idea seems to be well established that any old apple will do for cider making. This is far from true: it takes a good apple to make a good cider. A small number of dessert varieties were imported by the Normans and so it was not long before natural

cross-breeding occurred with the seedling trees 'over the orchard hedge'. The results of this haphazard crossing were the semi-wild apples that were probably the original cookers.

This may sound rather uncertain, but that is exactly what it is. At that time few people knew how to write, and by far the majority of those who could were the monks. Before the invention of printing everything had to be hand written, so there was a scarcity of information. It was not until printing arrived and with it the chance of mass communication that it was possible to spread the word far and wide on apple-growing techniques and varieties.

PEARMAIN AND COSTARD

By the end of the 13th century, a number of comparatively choice varieties were in existence, notably the Pearmain and the Costard.

The first mention of the English Pearmain was in a deed of 1204. Two hundred Pearmains a year, it seems, were given to the Crown for the lordship of the manor of Runham in Norfolk. The name Pearmain comes from the tapering (towards the eye), almost pear-like shape of the apple; there is a picture of it dated 1629 (see page 36). The fact that it was still prized 400 years after its introduction speaks well of this variety. Indeed, it remained popular until well into the 18th century, when it went into decline and eventually disappeared.

Where the Pearmain was the standard dessert apple, the Costard was its counterpart amongst cookers. One of the earliest references to this apple tells us that 100 were sold for a shilling in 1296. By 1325 its value appears to have risen to three shillings for 29. An inflation rate of something around tenfold in under 30 years seems a little excessive for those days.

The Costard was still the best cooker and very popular during Shakespeare's time but it disappeared early in the 18th century. It is, of course, from Costard apples that costermongers got their name. Unfortunately no Costard trees exist today but the early-season cooker Early Victoria (Emneth Early) comes pretty close to it.

PIPPINS

The next famous and worthwhile introduction does not appear to have come until about 1500 when the fruiterer to Henry VIII, Richard Harris, a famous man in the history of fruit, brought some scion shoots over from France. Amongst them were Pippins.

The name Pippin almost certainly refers to the fact that, unlike others, this particular type of cultivated apple comes true when propagated by seed. We saw earlier that hybrids produce only throwbacks when their seed is used for propagation purposes. However, given time, this can be overcome.

For example, the variety Bramley's Seedling has to be propagated vegetatively if it is to be true to type. If seeds are saved from a crop of Bramley apples and then sown, after five or six years we may be lucky and get a few seedlings that are something like the parent – along with many hundreds that are not. If the two most Bramley-like are crossed and the resulting seeds are sown, in a few more years there

will be more seedlings with Bramley characteristics. Again, the two that are the most like Bramley must be crossed and the seeds raised.

If this process is done often enough, you or your descendants will eventually reach a stage when virtually all the seedling trees are Bramley's. There is seldom complete uniformity, and you will still get the odd throwback, but it will normally be near enough for most of us. That, in fact, is the way that most of our vegetables and flowers arose and it is very probably how the Pippins originated.

Until the beginning of the 16th century apples were still being imported from France (*plus ça change, plus c'est la même chose*) but at last, with the arrival of the Pippins, the beginnings of a real apple industry in Britain were established. This was further encouraged when in 1575 farmer and fruit grower Leonard Mascall brought in some more Pippins and planted an orchard at Plumpton in Sussex.

Once Harris and Mascall were established, there was no holding them, and their wares, both fruit and trees, spread widely. In fact, it became something of a matter of honour amongst landowners to compete for the largest collection of apple varieties.

There were, however, variations amongst the Pippins and the keen eyes of these two growers were quick to spot any differences that might be to their advantage. One of the most famous of these seedlings was the Golden Pippin. This was 'held in high esteem' (a phrase that seems to be applied only to fruit) as a dessert and cider apple for several hundred years. It appears to have vanished without trace in the 1930s and, although there are many apples that masquerade under the same name, no tree is known to exist of the true and original Golden Pippin.

Men at work in a 17th-century orchard from New Orchard and Garden *by W. Lawson (1631)*

⌐ *Theories in apple* ⌐ *breeding*

One reason for its disappearance could have been its small fruit but Thomas Andrew Knight, a famous fruit grower who lived from the mid-18th to mid-19th century, believed that all varieties – not just individual trees – have a natural life span, towards the end of which they sink into a decline from which they never recover. What seems much more likely is that either the stock becomes affected by virus and, consequently, deteriorates or, as in modern times, the variety falls from favour and is replaced by a better one.

Knight's theory was absolutely serious but was based on the rather way-out belief that a graft or cutting, being part of an existing tree, would not live longer than the original tree, that is, somewhere around 100 years. The inference of this was that, when the original tree died of old age, its progeny would also die within a few years. It was an interesting theory and one which would have had devastating consequences had it turned out to be true. Can you imagine whole orchards of a particular variety dying off within a year or two of each other, no matter how old or young they were?

Fortunately, it was completely without foundation, but that was not known at the time. It therefore stimulated a genuine interest in breeding new varieties. After all, if they had a limited life, there would have to be a continual supply of new ones coming along to replace them.

Knight also believed firmly in seedling vigour and here he was much closer to the truth because seedling hybrids are known to be more vigorous than the seedlings of a true and direct line. He was the first person in Britain to breed apples deliberately by making intentional crosses between selected parents. As is so often the case, fortune never reflects the amount of hard work and enthusiasm that has gone into something because none of his seedlings was of any particular merit. However, his work did stimulate a far greater interest in and awareness of the whole subject.

Knight's experiments in breeding were not confined to apples by any means, but covered a complete range of fruit and vegetables. Probably his best known and certainly his most successful plant is the cherry Waterloo. It is still commercially grown and probably still the best black dessert variety.

⌐ *The choice widens* ⌐

The 16th and 17th centuries were certainly periods of enormous growth in the number of varieties. Besides those already mentioned, we find references to Apple-John, Baker Ditch, the French varieties Calville Blanc d'Hiver, Calville Blanc d'Eté and Calville Raoge d'Hiver, Catshead (illustrated on page 38), Codling, King of Apples, Leathercoat, Pomewater, Queening (also Queen of Apples) and Crimson Queening, Russeting, Summer Pearmain and Winter Pearmain.

OPPOSITE *Golden Pippins painted by William Hooker, from the* Herefordshire Pomona *(1876)*

Plate XXXVII

1. Golden Pippin

2. Scarlet G.P.

3. Franklins G.P.   4. Hughes G.P.   5. Pitmaston G.P.   6. Pine G.P.

Early apple varieties including (3) Pomewater, (5) Pearmain, (6) Queene Apple, (9) Kentish Codlin. From Paradisi in Sole *by John Parkinson (1629)*

Several of these still exist, not exactly in cultivation but in the apple variety collection at the National Fruit trials at Brogdale Experimental Horticulture Station at Faversham in Kent, England. The Queening and the Pearmain seem to have been the most popular. Apple-John (or John apple), Leathercoat and Russeting were well known for their keeping qualities, especially Russeting which would keep for up to a year.

A number of cooking apples also appeared at that time, notably the Codlin and Pomewater. Codlin was a name that originally appears to have referred to any small green apples that were used for cooking. In Elizabethan times, though, there was a distinct Codlin variety which was much prized for cooking and eating with cream; hence 'Codlins and cream'.

The Pomewater has disappeared altogether but was popular as an early to mid-season variety. Apothecaries favoured it for the manufacture of 'pomatum', or pomade.

A little later, a cooking apple called Harvey appeared which is still found today in many East Anglian gardens. This is named after a Dr Harvey of Cambridge and is mentioned in John Parkinson's book *Paradise in Sole Paradisus Terrestris* of 1629.

Cider varieties

Hitherto, cookers and wild crabs were the main raw ingredients of cider but a mixture of Pearmains and Pippins came into fashion again during the 16th and 17th centuries. This combination apparently gave a much milder brew. Crabs were also used on their own for making verjuice, widely used in Tudor and Elizabethan times for culinary and medicinal purposes.

Although cider making probably took place more or less all over the southern half of England to start with, it became concentrated in Hereford, Somerset and Devon and, oddly, in Norfolk and it is still based largely in these centres today.

With wild crab apples playing an important part in cider making it wasn't long before especially good ones were chosen, propagated and planted up in orchards. Bess Pool, for example, was found growing in a wood by a girl and was named after her. Mediate (Meadeate) was originally a tree that grew near a 'meadow gate' in Devon while Haglo Crab came from the village of Ecloe on the Severn. There was still no intentional breeding for new cider varieties but a lot of seedlings resulted from the practice of spreading the pulped and pressed apples (pomace) on the orchard floor, resulting in another by-product, no doubt – an interesting smell!

No one was more enthusiastic or famous for the improving of cider varieties than the Herefordshire landowner Lord Scudamore. He extended this interest into France when he was made the British Ambassador to Louis XIII and almost certainly collected scion wood from the famous Normandy cider orchards, some of which dated back to the 11th century.

From these, it seems reasonable to assume, originated many of the varieties that are still in use today. Several have the word 'Norman' in their name: for example, Cherry Norman and Strawberry Norman. With the increasing interest in apples of all sorts being shown by the landed gentry of the day, it wasn't long before the most enthusiastic among them were selecting and collecting new varieties themselves.

French apple varieties

Perhaps it was Lord Scudamore who started it, or perhaps not, but during Stuart times and the Restoration, a much greater interest was being shown in foreign varieties of apple, methods of growing and even in foreign gardeners.

John Evelyn, the diarist and author of *Sylva*, who just made it into the 18th century, was also a great champion of the French Calville varieties. However these apples enjoyed only limited popularity in Britain. They are a classic example of French varieties not thriving in England simply because they require a kinder climate. They were seldom grown commercially but succeeded well in sheltered gardens where the rather dwarf trees grew and cropped quite well if placed against a sunny wall. Exactly the same thing happened with Golden Delicious in the 1960s. Pomme d'Api seems to have had a similar reputation though it did better than the Calvilles because of its exceptionally good looks and delicate perfume. Indeed, it appears to have been grown for its appearance and scent rather than for consumption. Nonpareil seems to have enjoyed greater success and was commercially grown in Kent.

Development of the English apple

By the end of the 17th century, fruit growing was an established industry and it continued to expand throughout the 18th century. The number of varieties also increased but, as so often happens, it was the older ones that still dominated the scene, varieties such as Juneating (early dessert), Codlin, Pearmain, Golden Pippin, Russets, Queening, Catshead, Nonpareil and Pomme d'Api. Fruit farming and market gardening also increased and, although the majority of holdings were within easy reach of the large London market, improvements in transport meant that they could be further away but still supply the same markets.

Nurseries also began to spring up all over the place, offering an increasing number of varieties. Horticultural writers became a force to be reckoned with as they played their part, telling everyone how best to go about growing this variety and that. In spite of all this activity, there was still very little in the way of an organized breeding programme. Most new apples were chance seedlings found by gardeners employed on the large country estates. With luck, they might be passed to a local nurseryman who would grow them for a while to establish what value, if any, they had.

PAGE 38 TOP LEFT *Catshead;* TOP RIGHT *Ribston Pippin;* CENTRE LEFT *Blenheim Orange;* CENTRE RIGHT *Cox's Orange Pippin;* BOTTOM LEFT *Worcester Pearmain;* BOTTOM RIGHT *Bramley's Seedling* PAGE 39 *'March' from a* Catalogue of Fruits *by Nurseryman Robert Furber (1732)*

RIBSTON PIPPIN AND BLENHEIM ORANGE

One such variety was Ribston Pippin. This appeared in the gardens of Ribston Hall near Knaresborough in Yorkshire. It is believed that it was a seedling from a batch of seed brought over to Britain from Rouen in France in about 1688. It was not exactly an overnight success; in fact, it was not until 1785 that we find it listed by a nurseryman at the famous Brompton Park Nursery on what was then the outskirts of London.

Even then its progress was slow and only 25 trees were raised as stock for the first three years. After that, however, it rose steadily in popularity, until by 1847 some 2,500 trees were being raised annually. By the mid-1800s it had become recognized as one of the finest eating apples of all time and it is still listed by nearly all good nurserymen. Yet strangely, it did not receive the Royal Horticultural Society's Award of Merit until 1962.

Opinions agree in principle that the original tree lived for upwards of 200 years at Ribston Hall. However, Hogg in his *Fruit Manual* of 1884 lets us into the secret that it was actually blown down in 1810 but, after attention, lived on until 1835. A sucker arose from below ground which, it was felt, 'with proper care, may become a tree'. It did; it grew and flourished until 1928 when it too succumbed to a gale.

Another famous apple, Blenheim Orange, was discovered in about 1740 by a Mr Kempster of Woodstock, the nearest town to Blenheim Palace. Like the Ribston Pippin before it, though, it was many years before it was released for general use; 1818 to be exact.

COX'S ORANGE PIPPIN

The next milestone was one of the best things that ever happened to fruit growing and it occurred in about 1825. Living at Colnbrook near Slough was a retired brewer by the name of Richard Cox. He was an apple enthusiast and was in the habit of raising new trees from seed in the hope of finding something good.

He succeeded beyond all expectations. From two pips of a Ribston Pippin he grew the most famous British apple of all time, the Cox's Orange Pippin, and one that was almost as good, the cooker Cox's Pomona. In 1840 a Slough nursery, Smale and Sons, introduced these varieties to the public.

Cox's Orange Pippin was grown mainly in private gardens to start with. But, as fruit farmers became less suspicious of the new variety, its offspring spread along with its reputation so that, by 1883, it was one of the leading dessert varieties. It is now grown all over the world and its excellent flavour and appearance are still rated of the highest order.

The original tree is believed to have been blown down in 1911 but in 1933 there still stood in the same garden two venerable trees raised from grafts of the original. At that time, they were certainly the oldest Cox trees in existence. One reference suggests that one was Orange Pippin and the other Pomona.

BRAMLEY'S SEEDLING

The greatest cooker of them all, Bramley's Seedling, came on the scene at about the same time. The origin of the name is obscure but it sounds as though it was named after someone. One thing is certain: Mary Ann Brailsford raised it at Southwell, Nottinghamshire, between 1809 and 1813. After a long wait it was introduced in 1876 by Mr Merryweather, a nurseryman also of Southwell (he was also responsible for the widely grown Merryweather Damson).

WORCESTER PEARMAIN

The last variety to take its place in the apple Hall of Fame, for the moment, is Worcester Pearmain. It was a seedling from Devonshire Quarrenden and was introduced in 1873 by a Mr Smith, a nurseryman of Worcester. As an early eater, it was the main commercial variety for almost 100 years – until the variety Discovery came along in the 1960s, in fact.

Although the introduction of these famous varieties hardly brings us up to modern times, nothing approaching them in importance has appeared for nearly a century. Cox and Bramley are still the most popular and widely grown apples in Britain, both commercially and in gardens, and this over 180 years after their birth. It might be felt that this simply demonstrates the conservatism of the British but it also shows, undeniably, the excellence of the two varieties. And thanks to the methods of propagation by budding and grafting developed centuries before, these important varieties could be increased with ease and speed.

Exchange of ideas

It also happened at about that time that the, as it were, 'national' crops developed. Along with Britain, the other great European fruit-growing nation was France. However, fruit growing in France veered towards pears whereas in Britain it was apples that became the chief fruit crop. And today this national preference still exists. It is a fact of horticultural life that the climate in Britain favours apple growing and in France pears, so it comes as no surprise that the best apples are grown in Britain and the best pears in France.

Although the horticulturists of the two countries have indulged in a certain amount of friendly rivalry there has also been a massive exchange of ideas and practices over the years. This was sometimes more relevant to pears than to apples but the botanical similarity of the two meant that much of the information going back and forth across the Channel could be adapted to national needs and preferences.

III

Methods of Cultivation

In the development of apple growing in Britain, or rather in England – for there was very little in the other parts of the British Isles – there had always been a strong tie between the private and commercial sectors. In the 19th century there had been not only an international exchange of ideas but also one between the commercial sector and the private worlds. This was hardly surprising really as the commercial was often run by the large landowners, the very people who also grew fruit to perfection in their extensive gardens. It was as a result of the experience gained in private gardens that practices spread to the commercial orchards and, presumably to a lesser extent, vice versa.

Quality v. quantity

There was, however, one essential difference between the two. In the private gardens quality was the all-important factor whereas in the commercial orchards it was mainly quantity that mattered. To a great extent the difference was irreconcilable and, as the techniques of both systems improved, the two forms of cultivation grew further and further apart.

Thus, as the development of commercial orchards in the 19th century moved away from the country gardens and estates and into the hands of farmers and specialist fruit growers, the growing methods and practices developed along completely separate lines. The different ends justified the different means. This resulted in the two main systems that still exist today: 'extensive' apple growing in orchards and the 'intensive' methods that are used in gardens. But now there is a drift back towards the same type of low-cost, low-maintenance methods of cultivation.

OPPOSITE
In October *by Percy Tarrant (1883–1904)*

On paper, the difference between extensive and intensive methods is that extensive growing involves large trees with plenty of room between them in orchards, whereas intensive systems consist of much smaller trees and more concentrated planting, with the trees often trained and planted close together. Clearly, farm orchards have to be easy to maintain, with little call on labour except at picking time. Private gardeners, on the other hand, are much more concerned with excellence and can, therefore, afford to spend a lot more time and money on the production and maintenance of their trees.

One could be forgiven for thinking that this state of affairs led to chaos but it says a great deal for the adaptability of the apple that it did not. In fact, these two extremes existed happily side by side in the 19th century and well into the present one and it was only the First World War that really spelt the end of the great private gardens. Trees could, and still can, be either 12 m (40 ft) high and 12 m (40 ft) apart in an orchard or trained into intricate shapes and different sizes against walls and fences ideal for today's tiny modern gardens.

ᴪ *Rootstocks* ᴪ

The variety of apple will have a bearing on the eventual height of a tree. For example, Bramley is extremely vigorous and will quickly grow to a great height whereas Worcester Pearmain takes quite a long time to get anywhere. But it is perfectly possible to grow, say, two Cox's Orange Pippins, one 6 m (20 ft) high and the other 2 m (6 ft). This can be achieved as a by-product of budding and grafting scions onto rootstocks of different degrees of vigour.

Back in the mists of time, the drill for getting rootstocks was to go out into the woods in the winter and collect wild apple seedlings that had sprung up. These were taken back to the home or the nursery where they were grafted or budded the following spring and summer to produce new apple trees.

This was fine from the point of view of economy; you obtained as many rootstocks as you wanted and they cost you next to nothing. But there was one snag. The whole point of budding or grafting was to produce an infinite number of trees of the same variety, so avoiding seedling variability.

However, if seedling rootstocks were used for growing the new trees, variations in vigour would be automatically introduced. Whilst all the resulting trees would be of the same variety, they might be strong growers or weak growers. There was no way of telling how they were going to turn out until it was too late. And this is still true.

A system that was used quite widely for raising rootstocks for cider apple trees harks back to the practice of spreading the pomace (the crushed and pressed remains of the apples) beneath the trees as a kind of manure. Naturally enough this gave rise to enormous quantities of seedlings, but unfortunately these were all as variable as the wild ones.

An orchard ladder from Plan and Full Instructions to Raise all Sorts of Fruit Trees *by T. Langford (1696)*

An orchard with wall-trained fruit trees from The Clergyman's Recreation *by Lawrence (1715)*

THE SEARCH FOR UNIFORMITY

A refinement on this was still to raise the rootstocks from seed but to be sure that this came only from true species of crab; in that way, the rootstocks would all be virtually the same. By choosing the species of crab according to its vigour, a reasonable influence on the size of the tree that was being produced could be ensured. A vigorous crab would give rise to a large tree and a weaker one to a smaller tree.

STOOLING

1 *Established tree*

2 *Cut down after one year of growth*

3 *Earth up early in summer*

4 *Rooted shoots in winter can be removed and grown on for use as rootstocks*

It sounds simple and it is, in fact, the same principle by which tree vigour is controlled today. However, by modern standards, there is still an unacceptable degree of variation even amongst the selected seedlings. One way round the problem was to root cuttings from a crab apple of the required vigour and use these as rootstocks. Another was to dig up suckers from below suitable trees.

It is worth mentioning at this point why there appears to be this fetish of having consistently identical varieties and uniform rootstocks. The need for the varieties to be of uniform vigour is obvious but the rootstocks less so. Think then, what problems there would be if we had no way of knowing how vigorously a tree was going to grow and what size it was going to reach. This would be bad enough in gardens but just imagine a similar situation in an orchard. How far apart should the trees be planted? Will they need staking? When will they start cropping? Such unpredictability would result in absolute chaos.

Another trouble with those early rootstocks was that there was not a sufficiently dwarfing one to make it possible to grow small trees. Most ended up like cathedrals and were extremely difficult to manage. It was realized that if these virtues of predictable uniformity and dwarfness could be developed, it would be a major turning point in fruit growing in general and apple growing in particular.

The first thing to tackle was the uniformity. This was easy in one respect because the desired stocks could be propagated vegetatively. Thus a 'clone' would be created – a collection of plants reproduced vegetatively and all identical in every respect.

STOOLING

The simplest method of propagation with this in mind was to 'stool' them. This technique is still widely used today; it is an easy and trouble-free method.

The parent rootstock is planted in the winter and allowed to establish itself during the following summer. A year after planting it is cut down almost to ground level. A number of new shoots will grow from the stump in the following spring. Once these are about 15 cm (6 in) high, they are earthed up to half their height. This is done once or twice again during the summer until about 18 cm (7 in) of the lower stem is covered. Throughout the summer and early autumn roots will then be produced on the buried sections of stem.

In the winter, the earth is scraped away and the rooted shoots are cut from the parent stool. Those that are well rooted can be used for grafting in March and April or for budding in July. The others are planted in nursery rows and grown on for another year to develop a stronger root system.

THE PARADISE STOCKS

The first rootstock that was capable of producing an apple tree smaller than its natural size was called the Paradise stock. It originally came from France in the first half of the 16th century and

Harvesting apples in 1873. The apple trees were large and vigorous and the orchard floor was frequently grazed by geese or other livestock

PAGE 47 Apple tree in fruit with beehive. the latter were frequently placed in orchards to ensure good pollination

could have been a natural hybrid between *Malus pumila* and *M. silvestris*. We are told that about a hundred years later these Paradise stocks were being used quite widely in France for their famous cordons and espaliers.

Another French apple that ended up being more valuable as a rootstock than as a fruit was the Doucin, though this was more vigorous than the Paradise.

To give an idea of the dwarfing nature of the Paradise stock, trees could be planted 2.7 m (9 ft) apart, whereas on seedling ('Free') stocks they needed up to 7 m (23 ft). By the end of the 17th century these dwarfing Paradise stocks were in quite good supply.

There was a problem, though, and it was a serious one. How could you be sure that what one nursery called a Paradise stock was indeed the same thing as the one another nursery was selling as Paradise 100 miles away? You simply couldn't. Added to this, it would not be possible to achieve any national, let alone international, agreement and uniformity until there was some central and omnipotent body which had the power to enforce standards. That was a long way off.

In the meantime, people had to muddle along as best they could with rootstocks that went under all sorts of odd names which were really only of significance to the nursery who raised and used them. There were stocks like the Broad-leaved English Paradise, Holstein Doucin, Jaune de Metz and Kelziner's Ideal, as well as the already known Free (Crab), Doucin and Paradise. The whole thing became so completely out of control that eventually any rootstock which was raised vegetatively was called Paradise and those raised from seed were known as Crab.

The root of the trouble was that no matter how genuine and worthy a nurseryman might be others saw no reason to fall in with his way of thinking and methods of classification. Even the famous Rivers nursery at Sawbridgeworth had fallen into the same Paradise trap and had given this name to two of their *vigorous* rootstocks.

Nothing authoritative and independent was done until 1871 when Mr Barron, the Superintendent of the Royal Horticultural Society's gardens at Chiswick, set up an apple rootstock trial. This set out to compare, observe and classify eighteen rootstocks in general use from home and abroad. These were grown both as whole trees and grafted with a single variety so that their tree habit could be seen as well as the effect that they had on the variety when they were used as a rootstock.

The first thing that the trial confirmed was the great variation in what different nurseries called Paradise stocks. Barron, however, was pretty sure that he had found the original French clone. 'Precocious' is the word that immediately springs to mind in its meaning of 'early developing' when describing the true and original French Paradise stock. Trees grown on it come into bearing much sooner than they do on other rootstocks.

One would think that the results of a trial as well organized as this one would have been greeted with enthusiasm, particularly as it was under the auspices of the Royal Horticultural Society, the governing body of horticulture in Britain. This appears not to have been the case at all because insanity continued to reign in the rootstock world. Even in the early part of this century, books were still talking about 'good selections of the Paradise stock'. How could anyone possibly tell a good one from a bad one?

Not until during the First World War was anything seriously done about it. Then, the newly created East Malling (pronounced Morling) Research Station in Kent took on the Herculean labour of creating order out of chaos. We'll be hearing a lot more about them later.

Pruning

The history of pruning goes back to the beginning of organized and deliberate cultivation of apple trees. Once people started to plant them, they also had to cut off branches to maintain the shape of the tree. Pruning is another example of how the two types of fruit growing developed in two different directions, the commercial and the private.

Cordons and espaliers had long been grown in the gardens of large private estates and, obviously, these trees had to be regularly trained and pruned to keep them tidy. On the commercial side, little pruning was carried out apart from the removal of those parts of the trees that impeded other operations. The bill-hook was the standard pruning tool at this time and branches were simply hacked out with no thought as to the welfare of the tree.

At about the end of the 18th century, a lot of the commercial orchards were deteriorating to a point where they were hardly worth keeping. In some instances this was probably because they were approaching the end of their natural life, but this was certainly not always the case. So serious was the situation that awards were offered for the best suggestions as to how the trend of deterioration could be reversed.

One of the first British references to pruning as anything like an organized job came in 1757 when Thomas Hitt gave an account of grafting and pruning when he wrote *A Treatise of Fruit Trees*. It still took time for the technique to catch on, though, and it appears that the first fruit farmer to carry out meaningful pruning on a commercial scale was Thomas Skip Dyot Bucknall. In 1797 he describes a much more enlightened and scientific approach to orchard management in his book *The Orchardist*, detailing methods of pruning for the definite purpose of improving the health and efficiency of trees. Bucknall can, in all probability, be said to be the father of modern commercial pruning.

The adoption of a proper pruning technique had a very far-reaching effect on fruit growing because it was perfectly obvious that the deterioration of many orchards had merely been brought about as a result of neglect.

This is clear when we consider what happens to any fruit tree which is left alone and allowed to look after itself (a euphemism for 'neglected'). This remedial pruning would almost certainly have meant the removal of dead, dying, diseased, broken, out-of-place and crowded branches. The effect would have been to open up the trees to the air and sunlight, to remove many sources of infection

The 'perfect forme of a fruit tree' from New Orchard and Garden *by W. Lawson (1631)*

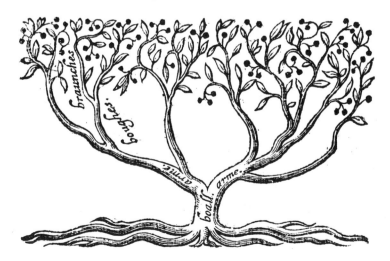

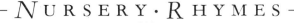

NURSERY · RHYMES

Apples are mentioned in a number of nursery rhymes and this is perhaps one of the most interesting examples, with its tune imitating a peel of bells. It obviously predates the present domed St Paul's Cathedral in London, which was built after the Great Fire of 1666. It also must have been a popular rhyme well before 1561 when the steeple of old St Paul's was destroyed by lightening.

The idea of an apple tree on the steeple is not as unlikely as it might sound as various spectacular and acrobatic events are recorded as having taken place on church steeples at that time.

and to encourage healthy new growth. In essence, it was what we now call the 'regulated' system of pruning.

The removal of diseased shoots and branches was particularly important; the fungus disease, apple canker, was rife in those early orchards. To this day, the surest way of controlling the disease is to cut off badly infected shoots and branches and burn them.

Very little thought went into those early attempts at pruning and they were performed almost entirely from the point of view of shaping the trees. Thus cordons and espaliers were trained into the desired shapes and pruning was largely confined to cutting off everything that was not either an extension growth or the formation of fruiting spurs. The time that such operations took was of little importance in gardens. What counted was the end product, neat and fruitful trees that were the envy of friends.

In commercial orchards, however, the production of high-quality fruits was not so necessary and, indeed, it is doubtful if the majority of farm workers would have been able to perform the relatively complicated task of pruning to a definite system. Little attention was therefore paid to the subject beyond what was essential to the health and welfare of the trees. It must be remembered that very few people solely grew fruit for a living. Most of them were engaged in mixed farming and depended on more than one product for their livelihood.

Short spur or 'three-bud system' pruning showing three different growth patterns (see also pages 53 and 54)

Autumn   *Winter*

Year 1

Year 2   *fruiting bud*

Year 3

'Honour'd Age' — at the end of the 18th century some trees were allowed to deteriorate to the point of hardly being worth keeping

Until the widespread use of the more dwarfing Paradise rootstock, it had been the general rule to grow large and widely spaced trees with sheep grazing the ground beneath them, and it was not until towards the end of the 18th century that methods of pruning were considered with a view to increasing the quantity and quality of the crops.

THE LORETTE SYSTEM

Although not necessarily the first to show an interest in pruning, a Frenchman called Louis Lorette was certainly the best known exponent of the technique and he had a great influence on the development of pruning. His methods were designed not only to shape and train the trees but to get the maximum weight of fruit from them at an early age. His work was connected mainly with pears but, is equally suitable for apples.

Lorette was the head gardener at the Wagnonville Agricultural College in north-east France and, it appears, was far too modest for his own good or he would probably have made a fortune. His work at the College started in the late 1890s but does not seem to have been noticed by anyone that mattered until about 1912 when Georges Truffaut came to hear of it. He was another power in French horticulture but a considerably more flamboyant character than Lorette.

Pruning apples and pears by the Lorette system involves no winter pruning at all; it is all done between June and September, whereas most other methods are carried out *only* in the winter. Unfortunately, the system is rather complicated and time consuming and takes place at a time of year when there are usually many other things to be done. However, many knowledgeable fruit growers have used the principle as a basis for simpler methods that are far more acceptable, the main one being the 'three-bud system'.

This is an all-winter system with no summer pruning and, by modern standards, is still quite complicated as there are many alternative ways of carrying it out. Putting it as simply as possible, all new shoots that have grown directly from a branch are cut back to three buds in the winter. By the following autumn, either none, one or two of the buds will have grown into shoots themselves. Those buds which have not will usually have formed short growths seldom more than 2.5 cm (1 in) long. What action is taken then depends on which course the tree took and is much easier described by means of a diagram.

This system is fine in gardens where there is no excessive pressure on time but there are obvious disadvantages in commercial orchards, the worst being the time it all takes. There is also the need to have properly trained staff. In spite of these difficulties, the commercial growing of cordons, complete with summer pruning, was quite widely practised as recently as the 1950s and 1960s but it largely disappeared in the 1970s when labour, land and time became too expensive.

STANDARD SYSTEMS

In its early days pruning methods were sharply divided between private gardens and commercial orchards and this difference continued throughout the 19th century and well into the 20th. It lasted, in fact, until East Malling tackled the problem of sorting out standard systems of pruning to suit pretty well every commercial situation. This resulted in the establishment of the Regulated system, which was suitable for the larger and unrestricted tree forms used in orchards. But certain varieties perform much better when they are closely pruned each winter, roughly along the lines of well-trained trees. For these, it was found that 'Spur' pruning was more suitable. Miller's Seedling and Early Victoria are two such varieties.

The tree consists of a series of permanent branches on which both temporary fruiting shoots and established fruiting spurs are allowed to form. Pruning is carried out in the winter when any strong new shoots not required for branches are cut out. Very weak new shoots are shortened back to about 8 cm (3 in) to form spurs. Shoots of medium vigour are either left to fruit, if they are not likely to be in the way, and then cut back to 8 cm (3 in) or are shortened to 8 cm (3 in) in the first place. New shoots arising from existing spurs are cut hard back to 2.5 cm (1 in). Easy to follow but still time-consuming.

⸙ *Pests and diseases* ⸙

The control of insects and other pests, as opposed to fungus diseases, seems to have played a comparatively minor part in the cultivation of apples until quite recently.

Although there are frequent references to pest control throughout gardening literature, it is only since the last quarter of the 19th century that it appears to have been deemed really necessary. The

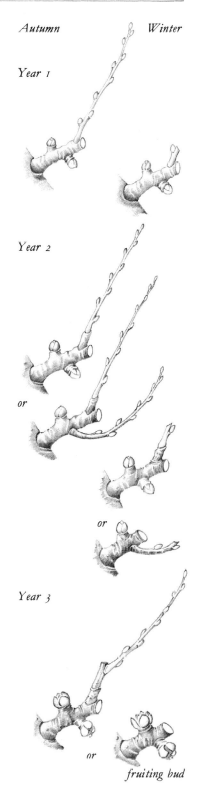

Autumn   *Winter*

Year 1

Year 2

or

or

or

Year 3

or

fruiting bud

Autumn   *Winter*

Year 1

Year 2

or

Year 3   *or*

                fruiting bud

        or

OPPOSITE *The apples in this
Still Life by Osias Beert the
Elder (1570–1624) show signs
of attack by pests*

reason for this could be that until the cultivation of fruit became more widespread and sophisticated, Nature seemed perfectly well able to keep things under control by itself and without any help from us. Obviously the cheapest and least troublesome way was to let the friendly pests and predators keep control of the damaging ones.

We know from old pictures of fruit that this was no more satisfactory then than it is now. The fruits were as covered in scars and bites from the attacks of pests as they are today when Nature is allowed to take its course. When, towards the end of the last century, the time came to do something about it. The chemicals available were just as dangerous to the operator as to the target. This was very likely another reason for the lack of enthusiasm.

PEST CONTROL MEASURES

The chronological written history of pest control is rather sparse when compared with the records of tackling disease, but we do know that by the end of the 19th century there were some positive recommendations. For example, the control of the codling moth grub, which is often found in ripe apples, involved the use of two vicious chemicals, either Paris Green (copper arsenite) or the rather similar and equally colourful London Purple. These were applied at the petal-fall stage and again at the suitably vague time of 'when the fruits begin to swell'. In fact, even by today's high standards these times correspond fairly accurately to the modern recommendations of the first and third weeks of June.

Another interesting point when considering the safety of chemicals is that until as recently as the 1960s lead arsenate was the recognised remedy for codling moth in commercial orchards.

The control of greenfly (aphids) was no less complicated. It consisted of making a solution of tobacco water (nicotine) with soft soap as a wetting agent. This could be made by steeping 500 g (1 lb) of tobacco in 18 litres (4 gal) of water and then adding 110 g (4 oz) of soft soap. For the anti-chemical gardeners of the day, the recommendation was to blast the creatures out of the trees with powerful jets of water and then rake the ground to kill those that had been washed off. Crude, but apparently effective.

Winter washing was also carried out but instead of tar oil or DNOC, the remedy early in the 20th century was a 2 per cent solution of caustic soda with soft soap added both as a wetting agent and to increase the number of victims.

CONTROL OF DISEASES

On the disease control scene we have much greater detail of its progress. You will remember that the first outside influence to create any real problems for apple trees was the fungus disease canker. It was this that really provided the impetus for pruning. The disease is common and serious to this day. It enters the tree through a wound, or even the scar left when a leaf or fruit falls. A rot sets in which, although mainly in the bark, will extend into the wood. Once the

lesion (wound) girdles a shoot or branch, the growth beyond it dies through lack of sap. Neglect and poor drainage are the main causes of canker but some varieties are also more susceptible to it than others. Cox, for instance, is especially bad.

The man who established that pruning could be the answer to tree deterioration, Thomas Bucknall, also recommended that a mercury paint be applied to any saw cuts made during pruning. This is one of the first mentions of mercury being an effective fungicide and some modern canker paints are still based on a similar formula to his.

LIME SULPHUR

Another early reference to the use of chemicals for controlling fungi dates from the mid-1840s, in Margate. Anyone living in Ireland at that time would have been only too painfully aware of the effect of fungus diseases. It was what we now call potato blight that caused the appalling potato famine: in the fifteen years from 1845 to 1860, a million people died in Ireland as a direct result of the famine and a further million and a half emigrated.

Is it any wonder, then, that the grape growers of Margate (there cannot have been too many of them), when faced with mildew for the first time, threw up their arms in despair and accepted it as the will of the Almighty? All, that is, except Mr Tucker, gardener to a Mr J. Slater.

In this respect he was a remarkable man because, instead of leaving everything to someone else and waiting for something to turn up, he set about looking for a remedy himself. He had been in the habit of applying flowers of sulphur to the peach trees in his care to combat mildew and decided to give it a try on the grapes as well. However, instead of using it as a dust, as he did on the peaches, for no apparent reason he mixed it with lime and water and applied it to the affected parts of the vine. It worked like a charm and was the original and rather crude form of lime sulphur, a fungicide which has only very recently ceased to be manufactured.

If this remedy was received gleefully by the vine growers of Margate, imagine how the French growers felt about it. It was the answer to their prayers and this was something that the British could say was their own invention.

France had not been idle over disease control. As long before as the 1760s, the Agricultural Society of Paris had been set up with the express purpose of finding a cure for stinking smut (bunt). The best remedy that they could come up with in the short term was steeping the seed grain in some witch's brew, or even plain water, before sowing it. Not very scientific but, if it worked, did that matter?

Unfortunately, it didn't work very well. Then Abbé Tessier tried treating the grain with almost everything he could lay his hands on, even including brandy and crême de menthe. None of his crude seed dips was particularly effective but he did establish one vital piece of information. Whether it was used in the dip or for mixing with the grain afterwards to dry it, the presence of lime was necessary for success. Could it possibly be the lime that was the beneficial agent and not the concoction itself?

The next person to carry on this work was Benedict Prevost. He discovered that fungus spores were living organisms; they had previously been regarded as mineral. With that established, at least it was now known what form the enemy took.

COPPER SULPHATE

His next discovery was of even greater importance and, as was so often the case, it came about by chance. He noticed that on a farm where wheat seed was put in copper baskets or sieves before steeping, the effect was immeasurably improved. In addition, when copper containers were used in his laboratory experiments, the results were chaotic because the fungus spores failed to germinate. At this point he realized what was happening: copper was lethal to fungi. After further experiments, he established that even one part per million of copper sulphate in water was enough to prevent bunt spores growing.

All this was happening early in the 19th century and, naturally, much of this information spread to England. However, it is interesting to note that whilst copper and lime were the materials attracting the most attention amongst growers in France, sulphur was of chief importance in Britain.

Both sulphur and copper were tricky materials where plants were concerned. Some species, and even varieties, would put up with them but others showed signs of damage. From the French experiences, copper appeared to be marginally the safer of the two.

BORDEAUX MIXTURE

The next big step forward with vines came in 1882 in France when it was noticed that plants beside the paths in a certain vineyard were still healthy and in full leaf whilst those further away had been defoliated by mildew. Upon investigation, it was found that the vines adjoining the paths had been treated with a mixture of copper sulphate and lime as a deterrent to 'scrumpers'. This concoction was the original and crude form of the popular fungicide Bordeaux mixture. In fact, it was in 1881 in France, just before the 'invention' of Bordeaux mixture, that Paul Olivier found that a weak solution of copper sulphate had definite fungicidal action against pear scab.

There soon followed work on apples in America using both copper sulphate and the newly found Bordeaux mixture. These trials established that Bordeaux could be used perfectly safely and effectively at a reduced rate for the control of scab, easily the most serious fungus disease of apples. This led to the first definitive recommendation for controlling a fungus disease on apples.

In 1895 the word went out from America that four sprays of Bordeaux mixture would prevent scab attacking apples. The first should be applied when the buds opened; the second, just before blossom; the third, just after the petals had fallen; and the last when the fruitlets were 1 cm ($\frac{1}{2}$ in) across. That advice still holds good today, even though there are more sophisticated and safer fungicides available, such as benomyl.

Blossom stages for spraying against scab

bud opening (bud burst)

just before blossom opens (pink or white blossom)

petal fall

fruitlet

I V

Apple Growing in Modern Times

y far the greatest leap forward in the development of apple cultivation was the establishment of reliable methods of vegetative propagation, and this occurred quite early in the story when budding and grafting replaced the hit-or-miss approach of growing new trees from seed and, later, growing them from cuttings. At last it was possible to grow trees that would all be the same and, ultimately, would be of a predictable size and vigour.

At the time that the East Malling Research Station was set up near Maidstone in Kent just before the First World War the rootstock situation was chaotic with the same names, such as Paradise, referring to several different stocks of widely differing vigour. The original Paradise, you will remember, was a dwarfing rootstock from France.

One of East Malling's first jobs, therefore, was to sort the whole lot out and introduce a standard range that everyone would know and use. This daunting task was undertaken by the Station's Director, George Hatton, who was later knighted.

Classification of rootstocks

All the rootstocks known to exist in English nurseries and which it was felt were worth further investigation were gathered together. They were sorted out, classified and eventually numbered and reissued to the nurseries as standard clones of true and selected rootstocks. Because the demand was mainly for a dwarfing stock, not surprisingly the first to be issued was the English Broad-leaved Paradise. However, rather than retaining the existing name, which

OPPOSITE
Apple Gatherers *by Stanley Spencer (1891–1959)*

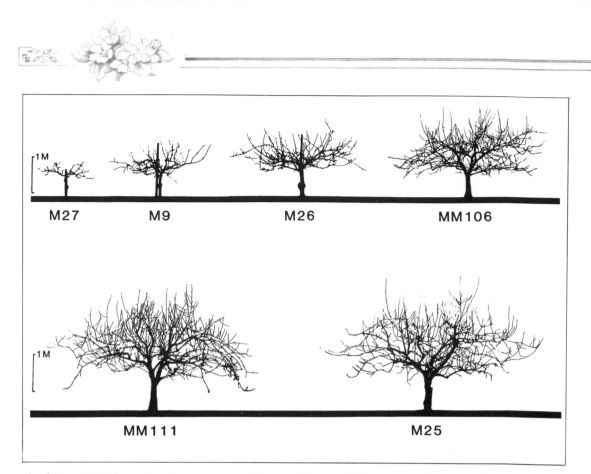

M27 M9 M26 MM106

MM111 M25

A selection of widely used apple rootstocks showing varying degrees of vigour

meant different things to different people, it was issued with the somewhat uninspiring but absolutely clear title of Malling Type 1, and that is what it is to this day, shortened to M1.

These original rootstock clones were given Roman numerals but all have now been changed to Arabic to avoid confusion. M1 is little used today, however. Cox does not perform well on it and anything that has that reputation is unlikely to be popular when there are plenty of alternatives.

This re-selecting and re-cloning continued and now we have reached M27, though naturally not all twenty-seven Malling rootstocks are still in use. Many have been replaced by improvements and forgotten. Some ten can still be found but, of these, only about four are in common use, the most popular being M9 and M26.

MALLING MERTON ROOTSTOCKS

Another research station came into the picture when the importance of apple rootstocks was realized. This was the John Innes Horticultural Institution, originally at Merton. Collaboration between the two stations was very successful and resulted in the Malling Merton rootstocks. To identify these as a completely new generation, the new ones started at 100 and were prefixed MM, for example MM 106.

All seemed to be going well until, from time to time, it appeared as though some of them were not behaving exactly as expected. It was hard to put a finger on it: it was just that some trees were different, and they shouldn't have been. After further research and consultation, it was discovered that the odd ones out among the trees were suffering from plant virus diseases, and that was what was changing their character. Although not exactly back to the drawing board, it did represent a setback, but work was soon in hand to clean up the stocks by destroying the virus. They were then reissued as virus-tested stocks free of all known viruses. The viruses had to be described as 'known' because it was modestly realized that there could still be viruses present that had not been isolated and identified.

THE EMLA RANGE

Work still continues on the virus testing and a new race of stocks is now available to nurserymen and commercial fruit growers called the E M L A range. These are just the same as they always were as regards vigour but collaborative work between East Malling and a Research station near Bristol, Long Ashton, resulted in new virus-tree clones, such as E M L A 9. The John Innes Institution had more or less ceased working on fruit during the 1960s, whereas Long Ashton had always been involved with it and still was. hence the tie-up with East Malling for the E M L A venture.

TYPES OF ROOTSTOCK

A look at the apple rootstocks available today and in general use is interesting as it shows what enormous variations are possible in trees of the same variety but grown on different stocks.

Rootstocks are classified into four broad types: dwarfing, semi-dwarfing, vigorous and very vigorous. M 9 and M 2 7 are dwarfing (M 2 7 very much so); M 2 6 and M M 1 0 6 are semi-dwarfing; M 2 and M M 1 1 1 are vigorous; and M 2 5 is very vigorous. All these are in reasonably common use and all, it is important to remember, have their virtues and vices.

M 9 is one of the oldest and is still very popular with both commercial growers and amateurs. It has been re-cloned, cleansed of virus and re-issued as EMLA9. It produces a small tree that comes into bearing early in life. The fruit is on the large side and usually ripens sooner than on more vigorous trees. This can be useful for growing early varieties when, for a commercial grower, every day can count. Probably the only big snag is that the roots are brittle, so the trees need staking or given some other support, for their whole life. Although an excellent rootstock on good soil, it sometimes has too little vigour on poor soils and can therefore give rise to rather weak trees. It is especially useful when growing cordons and other intensive systems but it is also excellent for apple trees to be grown in containers.

M 2 7 is a new and very interesting stock as it produces the smallest tree of all; hardly a bonsai, but not far off. It makes a good dwarf tree and is perfect for use in containers, though not to be recommended for poor soils. Like M 9, it needs permanent support.

The apple variety Kent growing on M9 rootstock. M9 encourages early bearing and large fruit but needs a good soil to perform well

M 2 6 is larger in size and is described as semi-dwarfing. It still produces a small tree and is much used commercially with modern orchards where lack of size is important. In fact, it is the result of a cross between M 9 and the very vigorous but now redundant M 1 6. It is better than M 9 on poor soils but it also needs staking, and it is more tolerant of mineral deficiencies in the soil than other stocks. It does, however, have a tendency to start slowly and poorly, so give it a chance and ensure that it has good planting conditions.

M M 1 0 6 is a very popular and widely used rootstock, particularly with commercial growers using intensive methods or growing bush trees. It produces a medium-sized tree, bigger than M 2 6. Trees crop rapidly and are reasonably happy on poorish land, which makes them especially suitable for, and popular with, amateurs. They sometimes overcrop and produce poor-sized fruit but that can be cured by thinning the fruit.

M 2 used to be very popular for bush trees but has now given way to the more modern M M 1 0 6 and M M 1 1 1. It was a popular rootstock in nurseries because it could be propagated easily. It was one of the main standard commercial rootstocks until the M M range was introduced.

M M 1 1 1 has largely replaced M 2 on account of its heavier crops, and is now regarded as the best of the vigorous rootstocks. It is resistant to certain mineral deficiencies, notably potash and magnesium, but the greatest advantage of this rootstock is its ability to withstand drought conditions.

M 2 5 is the most vigorous rootstock we are likely to be confronted with. It has replaced M 1 6 because trees on it crop at a younger age, and they also show resistance to the woolly aphid pest.

These are the main rootstocks likely to be encountered and, clearly, there is a wide range to choose from. The saving grace for the amateur is that there is seldom such a wide choice in the nursery, since either M9, M26 or MM106 will answer virtually all circumstances and conditions.

DOUBLE WORKING

With the tendency for dwarf, or at least small, trees to be the most popular in orchards and gardens, it follows that the least vigorous rootstocks, M9 and M26, are likely to be the most widely used. However, it can be seen that they have the same snag: they both need permanent staking. How nice it would be if there were some way of overcoming this. The trouble is that, every time a dwarfing rootstock comes along, it has the same fault – a weak root system.

The problem was eventually overcome in quite a simple way. Instead of the dwarfing M9 being the rootstock, another is used. On to that is budded or grafted a short length of M9 and on to that is worked the variety. Thus a Cox tree, for example, could be growing on, say, MM111 but with an M9 'interstock'. This gives a tree that is well anchored on an excellent stock but with the vigour of M9. It's as easy as that.

This double working, as it is called, was used extensively in the past with the pear variety Williams; it still is, come to that. The reason is that Williams is not compatible with quince rootstocks, which are commonly used for pears. Other varieties of pear are compatible, so a sliver of one is worked on to the quince stock and the Williams is grafted on to that. This is the same principle as that of the apple interstock. As always, there is a snag but it is not a cultural one this time: it is simply that trees which have been double worked are bound to cost more than the ordinary ones. But how many difficulties are overcome at no cost at all?

ABOVE Malus '*Profusion*' *growing on the very dwarfing* M27 *stock.*

Grafting and budding

In many respects the phenomenon of grafting and budding is one of the most versatile connected with apple growing. Not only does it enable us to plant trees of known and predictable vigour but it also overcomes difficulties such as incompatibility. However, if this ability of one bit of wood to fuse and grow on to another is successful at ground level, it must follow that it can also operate higher up in the tree itself.

FAMILY TREES

One use of this facility that is especially valuable in gardens is what are called 'Family' trees. These are single trees but composed of more than one variety. What happens is that the nurseryman chooses, say, a Cox tree and then grafts on to one of its branches another variety, Discovery for instance. Thus a single tree will have two varieties of fruit on it. In most cases, the nurseryman will graft on two others so

BELOW *An example of double working. Here Cox is growing on* MM111 *with* M9 *interstock*

that family trees usually contain three varieties. This is especially convenient for small gardens because there is a choice of varieties without the loss of space associated with three trees.

In commercial orchards there is a more important use for grafting and that is in order to change over from one variety to another on a tree. There can be a number of reasons why a fruit farmer should want to be rid of one variety and replace it with another but, whatever they are, grafting is often the cheapest and most convenient way of doing it.

For a start, it means that the existing trees can be retained and used, if you like, as large rootstocks. It avoids the cost of grubbing up the orchard and replanting it with new trees and it means that profitable crops will be produced relatively quickly and certainly sooner than on new trees. Obviously the trees have to be young and healthy enough to make the operation viable but that is seldom a problem.

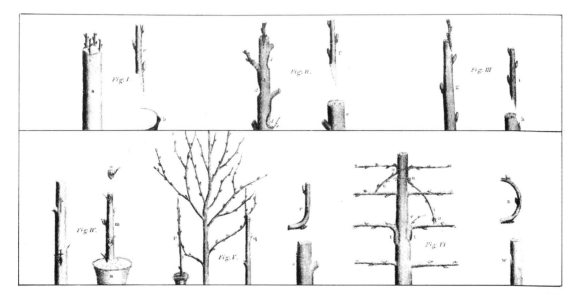

An illustration from Treatise on the Cultivation and Management of Fruit Trees *by W. Forsyth (1797), showing types of cleft grafts, budding and approach grafts*

METHODS OF GRAFTING

The art of grafting was discovered and established hundreds of years ago and we know from pictures that the different types of graft are largely the same today. But what are these apparently magical operations that we perform which enable a bit of wood to grow into a tree? To understand what goes on and what has to be done really requires a simple botany lesson.

Between the bark and the wood of a tree there is a layer of cells called the cambium. These cells are constantly dividing and multiplying and are responsible for making new wood. How well they operate can be seen by examining the width of the annual rings in the cross-section of a branch. For a graft to be successful, the cambium layers of the branch and of the shoot (scion) that is going to be grafted to it must be in contact over as large an area as possible.

The more cambium there is touching, the greater the success of the graft and the quicker it will take. The second essential is that the area of the graft (the union) must be kept completely airtight. Any drying out is likely to result in the failure of the graft.

The best time of the year for grafting is when the sap is starting to rise in the wood again, usually in April. Not only does the moving sap allow the bark to be lifted easily from the wood but the weather is also warming up then, so the union starts to form quickly. It has to be done in the dormant season when there are no leaves on the scion but, in any case, you would not want to pick the freezing weather of February or March when you are trying to be efficient and businesslike with a razor-sharp knife and numb fingers.

Before the actual grafting can take place, a certain amount of preparation has to be gone through. As only dormant wood of the new scion variety is suitable, one-year-old shoots of the required variety have to be gathered and heeled into the ground during winter pruning so that they remain dormant until April.

Top-worked trees using a number of cleft grafts from The Booke of Arte and Maner, Howe to Plant and Graffe all Sortes of Trees *by Leonard Mascall (1572)*

TYPES OF GRAFT

There are a number of different types of graft used. Some require just the bark to be lifted and the specially prepared scion to be inserted under it. These are descriptively called 'rind grafts'. A similar, but initially stronger, graft requires the stump to be split across the top as though splitting a log, and into this split is pushed the scion. Understandably, this is called a 'cleft graft' and the scion is held in place with considerable force.

Two basic grafting systems are employed when working established trees over to another variety. The normal one is called 'top working' or 'crown grafting'; they are simply different names for the same thing. With this, the main branches of the tree in question are cut back in the winter to about 60 cm (2 ft) from their point of origin at the crotch of the tree. It is on to these stumps that the scions are later grafted. This method uses less grafts per tree but it does take longer for the new tree to grow and develop.

With the other system, known as 'frame working', only the small shoots and branches are removed and the grafts are placed along the length of the retained branches. This takes longer to perform and requires a greater number of scions and different types of graft but it does have the advantage of forming a new tree considerably sooner.

Work on preparing for grafting starts in the winter and, because of the magnitude of the operation with top working and the consequent shock to the tree's system, it is sometimes spread over two years, especially with bigger trees. Half the branches are removed in the first winter and the stumps grafted, while the others are left for the following year.

One point is important, whichever timing is adopted. All the shoots and small branches growing on the stumps must be left on for a further two to three years to act as 'sap drawers'. The purpose of this is to produce leaves which will then draw up the sap, without which the grafts are less likely to be a success. If the initial cutting back is done in the early winter, many grafters leave the stumps longer than necessary and cut them back by the required amount immediately before grafting. This ensures that the site of the graft is healthy and living.

After any graft, all cut surfaces are treated with a special grafting wax to keep out air and fungal infections. With the rind grafts, the areas of the union have to be taped to hold the scions in place before waxing. As one might imagine, it is skilled work but, in the hands of someone who knows what they are doing, the success rate can be over 95 per cent.

Top-worked tree

Frame-worked tree
(arrows indicate some of the more prominent graft unions)

Cuttings

In spite of that, and especially where the raising of young trees is concerned, budding and grafting are expensive and lengthy jobs and this is reflected in the ultimate cost of the trees to growers and gardeners. Research, therefore, is now being directed towards a system that would enable the trees to be raised by cheaper means, primarily from cuttings. There have already been successes with raising rootstocks in this way, so it should be only a short time before we see consistent results with apple varieties.

At present the best method lies in the use of hardwood cuttings taken during the early winter and rooted in centrally heated boxes of rooting compost. Once roots begin to appear, the cuttings are weaned and planted outside in nursery rows in the same way as they are from stool beds.

*A*PPLE·*B*OBBING

This is a traditional Hallowe'en game, particularly popular in America. Children and grown-ups bend over bowls of water to try and catch an apple, hands are definitely not allowed! It is an old, old game and its origins are believed to go back to an ancient harvest rite, perhaps to honour Pomona, Roman goddess of fruit, or perhaps to one carried out by druids.

This system is showing promise but there are many problems still to be overcome before it can be recommended, even for specialized nurseries. One of the worst is that, as with ornamental trees and shrubs, different varieties vary in their success rates. This applies to both rootstocks and scion varieties. Some will root relatively easily whereas others are extremely difficult. This obviously has to be overcome or a situation would exist where some are being raised from cuttings and others from budding or grafting.

Another thing that could be troublesome is that one of the reasons for using rootstocks is to control the size and vigour of the apple tree. If the rootstock is done away with by rooting cuttings of the variety, surely we will be losing one advantage in order to attain another. In fact, Coxes are being grown on their own roots in trials at research stations now but it is still too early to come to any firm conclusions. So far, though, they look good.

Micropropagation of rootstocks

Another possibility being explored is the use of micropropagation to raise both scion varieties and rootstocks. This is a fascinating subject and involves rooting minute pieces of plant tissue in nutrient jellies. Before examining the way in which the operation is carried out, it would be helpful to look at the benefits, or otherwise, of the system.

The idea was born and the first experiments carried out as long ago as the 1830s but it was not for about another hundred years that scientists considered micropropagation as anything more than a good idea which was sometimes successful but often not. Between about 1940 and 1980 there was a rapid build-up of interest which was accompanied by an improvement in techniques and an expansion in the number of plants that could be propagated in this way. Today there are several commercial firms in Britain specializing in micropropagation and many others throughout the world.

THE BENEFITS

One of the original benefits was in the production of virus-free fruit plants. It had been discovered that in most infected plants there is one tiny part of them that remains free of virus. This is a minute portion at the very apex of the main stem – the growing point. Because a virus is a solid object that moves about in the sap stream of the host plant, it takes a little time for it to reach into newly formed plant tissue, and it was found that if this tip is used for propagation purposes, the resulting plants will all be virus-free. All that remained for the early scientists to find was a reliable way of using this micro-cutting.

In essence, this is what micropropagation means: the use of far smaller than normal pieces of plant material for propagation purposes.

As well as producing disease-free plants, the system has several other benefits. Because it is a vegetative method (as opposed to growing plants from seed), the progeny are essentially the same. A large number of new and identical plants can therefore be produced from just one original.

This has obvious advantages when building up a stock of a new variety. The more plants that can be grown and distributed, the sooner a working number will be available for sale on a general basis. Another benefit is that micropropagation is often successful where traditional methods are not. This is not always the case but it is more often than not.

On a rather different note, the ability to grow new plants from minute pieces makes it much easier to spread the newcomer far and wide. The need to distribute whole plants or large cuttings no longer exists. Another benefit, certainly with apple rootstocks, is that it can be carried on all the year round. Parent plants growing in pots can be used to produce the original cuttings even in winter, simply by bringing them into a glasshouse.

HOW IT IS DONE

There are about five different systems that can be employed but, for one reason or another, only two are in common use with fruit plants. One uses the growing point of the main stem and secondary shoots. The other uses callus tissue. This latter system is rather unreliable for the production of identical plants. There is a certain amount of variation possible which is not always seen at a stage early enough to correct. In the past this has led to problems with, amongst others, strawberry plants, but it is now well known and so is guarded against.

A word of warning here: micropropagation is certainly not the sort of thing that should be attempted without the right knowledge, equipment and conditions. It is a skilled and delicate operation which has to be done in laboratory conditions. Although connected with plants, that is as close as it comes to gardening. Also, it must be performed under the right conditions of temperature and light and in completely sterile surroundings.

The best system for propagating apple rootstocks uses the growing points. It is a three-stage operation in which, first, three 4-cm ($1\frac{1}{2}$-in) long tips are taken from growing rootstock plants. These are surface sterilized in diluted bleach. The top 1 to 2 cm ($\frac{1}{2}$ to $\frac{3}{4}$ in) of the shoots are removed and placed in culture tubes containing a nutrient jelly. After five to six weeks most will have become established and many will have started to callus. That is the point at which they are moved into a larger receptacle for the multiplication stage. During this time, new shoots will grow from the callus at the base of the original cutting. The number of new shoots will vary from two to six per month, depending on which rootstock is being propagated. On average, there will be about four.

Up to this point roots are not required. The emphasis has been on the production of shoots for multiplication, and this is accomplished by using a nutrient jelly containing shoot-producing hormones.

Once the stage is reached when the tiny new shoots are ready to be rooted, the cultures are removed to another container where a different nutrient jelly has been prepared. This one contains the right mixture of hormones to induce rooting.

It sounds complicated and it is but, when it became an established practice, it proved to be an extremely useful method of propagation. As with all new systems it had its drawbacks and was found after a time not to be the answer to all the nurserymen's prayers but it certainly answered a good many of them and a few leading nurseries have now set up their own units.

Creating new varieties

Of course, talk of improved propagation techniques and methods is all very fine but you must have something worth propagating in the first place and rootstocks are only one part of an apple tree. What most of us regard as an apple tree is simply the part above the ground, that is the variety and certainly not the roots. There is, therefore, always a good demand for new varieties. But here we come to a sticking point, for what is regarded as a good variety by one person might be useless to another. So many interests are involved and taste is such a personal thing that there cannot be an 'ideal' apple.

The most important factor is found right at the end of the chain – the consumer. If it is to have any future at all an apple must obviously be acceptable to the person who is going to eat it. How easy it is to grow or how well it crops is of secondary importance. If it does not have a reasonable taste, it is doomed.

The ease with which a variety can be grown is the next most important point, however. If it is hard to establish or if it takes a long time to come into bearing, commercial growers are unlikely to want it. Similarly, if it catches all the pests and diseases under the sun, no grower is going to bother with it. Those are the most obvious points by which a potential variety is judged but we then have to move into slightly more detailed areas.

The original Discovery tree grown from a seedling found in Essex

For example, if a new variety crops at the same time as an already well-established and excellent one, it is hardly likely to make an impression. Unless it has some obvious advantages. Also, if its appearance does it less than credit its chances of succeeding are slim, no matter how good it is in other respects. This is covered by the rather pompous description of 'eye appeal'. Conversely, if an apple is attractive to look at, its chances of succeeding on the open market are virtually guaranteed, whatever it tastes like.

Many other factors, too numerous to go into, also have to be considered in the search for new varieties and, as the years go by and more and more appear, the chances of finding real winners are progressively less. An interesting slant on the present-day position is that if a Cox or a Bramley had to compete with the standards laid down for a potential new variety neither would be acceptable: they have too many serious faults. It would be interesting to speculate how many other would-be world-beaters have been rejected along the way.

Serious apple breeding and all the complexities of selecting new varieties have a comparatively short history. Whereas the selection of naturally occurring and chance seedlings has been going on ever since apples were first cultivated, the deliberate breeding of new varieties with more than luck in mind is restricted to the last hundred years or so. It is only in this period that the rules of genetics have been understood beyond the point when it was realized that a red flower crossed with a white flower usually resulted in a pink flower.

SEEDLINGS AND MUTANTS

New apples still come from only two sources: professional plant breeders and observant or skilled amateurs. In the early days, most were the result of someone spotting a good seedling in someone's garden (Bramley for instance) but some were the result of actually trying to find something new (for example, Cox). Although this can still happen, it is becoming increasingly rare. The last important variety to come from an amateur source was the early dessert variety Discovery. This started life in Essex as recently as the late 1940s and was introduced into cultivation in the early 1960s.

Apart from seedlings, the other starting point is the mutant, or sport. There are many examples of these and, normally, the new one is simply a better colour than the parent. Many of the Cox sports come into this category. One of the most recently introduced is Norfolk Royal Russet, an improvement in many respects on the original Norfolk Royal.

In the mid-19th century an award scheme was instituted by the Royal Horticultural Society. This was intended to encourage the breeding of new apples and, inevitably, it led to a tidal wave of introductions, most of them completely useless. Eventually it became evident that it was the professional nurserymen with whom the power lay, people like Laxtons of Bedford, and for many years they were the main sources of new varieties in much the same way as new products are the life-blood of manufacturing companies today.

However, with the need for increased output and profitability, the emphasis moved away from breeding work and was directed into production. After all, it was the number of trees that a nursery sold that governed its efficiency, not the amount of behind-the-scenes work that went on. It was no surprise therefore when the newly set-up fruit research stations took on the job of breeding new varieties. They were dealing with the rootstocks so why not the varieties as well?

ABOVE *Merton Knave was a product of specialist breeding at the Johns Innes Institute*
BELOW *Norfolk Royal Russet started life as a sport or mutant*

SPECIALIST BREEDERS

East Malling and John Innes were again the leaders in the field. Anything with the word Merton in the name was a product of John Innes, such as Merton Charm, Merton Knave and Merton Worcester. Inevitably though, because it was concerned solely with fruit, East Malling stole the show with new varieties. It soon became the centre of the apple breeding programme and, indeed, has remained so ever since; though in rather a watered-down form because of reduced government backing. One wonders if we have seen the last of new English apples or if the mantle will fall somewhere else.

One of the most famous and successful apple breeders at East Malling was H. M. Tydeman. During the 1920s and 1930s he worked ceaselessly. Two of his best-known introductions were Tydeman's Early (originally Tydeman's Early Worcester) and Tydeman's Late Orange. Although not widely grown now, they were both important commercial varieties in their time, especially the former. Other important apple varieties to come out of East Malling include Michaelmas Red, Kent, Suntan, Jester, Greensleeves and the very new cooker, Bountiful. A new dessert variety, Fiesta, is due to be released in the late 1980s. This variety keeps well for an exceptionally long time.

Originally, it was the responsibility of the breeders to market their own introductions but for the state-aided research stations this proved to be rather a chore. None of them could afford to maintain a sales and marketing department of their own, so the distribution problem was often given to someone who had little knowledge of marketing.

With the introduction of Plant Breeder's Rights in 1964, the weight and complexity of the paperwork was beyond belief. In 1967 it was therefore decided to set up a central and separate sales and marketing organization to develop and distribute all new horticultural and agricultural plant varieties bred at the state-aided research stations. This was the National Seed Development Organization near Cambridge. For nearly twenty years the N.S.D.O. has carried out this valuable work of developing, building up stocks and selling home-bred apple varieties here and in 22 overseas countries. At the moment (1986) it is awaiting privatization.

CURRENT AIMS

Originally, the sort of things that plant breeders were looking for were the appearance of the fruit, its taste and texture, cropping period, length of storage life and all the other things that would make it a good variety for consumers and commercial growers to adopt. This even extended to the shape and vigour of the tree.

Although these factors are still important, the emphasis has now moved away from the more tangible and identifiable characteristics. What is currently most sought after is resistance, or even immunity, to pests and diseases. The main reason for this is to reduce the cost of production by not having to spray so frequently, but another important consideration is the public's increasing dislike of chemical sprays.

Breeding work is now including the use of some of the original wild *Malus* species that are known to carry genes which render them less susceptible to the disfiguring fungus diseases, scab and mildew. Transferring these resistant characteristics to new varieties is in everyone's interest, but it is one thing to create an apple with resistance to scab and quite another to breed one that is also palatable. And so the work continues.

Irradiation

Yet another very interesting technique for the specialist research station is the irradiation of existing varieties to create mutants. This may sound rather futuristic but all that the researchers are trying to do is to manipulate the genes of the apple variety to produce something better – a simple form of genetic engineering that can have nothing but good results. The work was started some years ago at Long Ashton Research Station but was transferred to East Malling when that became the national headquarters for apple research.

Incidentally, there is no danger of radiation or anything like that being imparted into the new variety. All that radiation does is to alter the characteristics of the chromosomes in the same way that the chemical, colchicine, was found to do many years ago. Much of the work at the moment is being carried out on Bramley's Seedling, the main commercial cooking variety, in the hope of reducing the vigour of the tree without affecting the quality of the fruit. Sexual cross-breeding produces a completely different apple. Altering certain genes will simply change some characteristics without destroying others. Cox, Discovery and Spartan are also being worked upon.

THE METHOD

One-year-old shoots of the variety in question are exposed to gamma radiation from a Cobalt 60 source. This has to be metered and monitored very carefully so that the result is mutation and not death. Too much radiation will kill the wood, too little will have no effect. Finding the right level of radiation is made all the harder by the fact that varieties differ in the amount they can stand, so this has to be established as well, usually by trial and error.

Following radiation, the treated shoots are made into grafts which are then worked on to ordinary apple rootstocks. The young trees cannot be used for assessment because the radiation has side effects which mask any true and lasting variations that might have occurred. During the following spring and early summer, new shoots are produced on the grafts and these are used to provide buds for budding in July. The resulting trees come 'true' and it is these that are used for trials. They are judged in the next growing season and any that are showing desirable characteristics are further propagated by budding. In this way a sufficient number of identical trees are created, cloned for replicated trials.

PAGES 74–5 Apple Gathering *by Frederick Morgan (1856–1927)*

During the first four years or so, assessment is purely on the basis of vigour and growth. Flowering and fruiting have to be left until the trees are older. Those that pass the growth test are later judged for flowering and fruiting and only after a few more years under the watchful eyes of the research staff are they finally released for trial on selected commercial fruit farms. If all goes well there, a few more years must still elapse before a sufficient number of trees can be built up for general distribution.

Altogether this is a very exciting development and one that is likely to produce the desired results far more quickly than traditional breeding. We already have the basic varieties that we want. All that now needs to be done is to iron out a few minor flaws that stop them being perfect.

The National Fruit Trials

One organization that must be mentioned is Brogdale Experimental Horticulture Station near Faversham in Kent. This is the home of the National Fruit Trials.

It is Brogdale's job to test and assess all new varieties that are considered to be worth the time and trouble. It caters for all types of commercial fruit hardy in the British Isles, not only apples. In addition, it houses the National Apple Collection which contains at least one tree of over 2,000 distinct varieties. These are kept to help research workers, students and anyone else with a legitimate interest in the identification of apple varieties but the main purpose of the collection is to provide a source of different varieties for breeding purposes. It forms, if you like, a gene bank of apple varieties old and new which can be called upon and used in breeding.

The characteristics of each variety are well known and documented so, for example, if a variety resistant to a particular disease is wanted, the staff will know exactly which to recommend. Whilst Brogdale is the national headquarters for testing new varieties, other Experimental Horticulture Stations around the country are used as regional bases for testing them in different climates and soils.

Tree forms and pruning

The enormous advances in apple rootstocks and varieties that have been made in the last 50 years have meant that the whole face of apple growing has changed. This has obviously been prompted by the needs of the commercial grower but, in many cases, the new techniques have been taken up enthusiastically by the amateur gardener. Not the least important of these modernizations has been the different shape of the trees that we now see in orchards and gardens.

Fifty years ago, the majority of commercial orchards were planted with 'standard' trees 9 to 12 m (30 to 40 ft) apart, giving as few as 25

Winged pyramid

to 50 trees per acre. This system was perfectly satisfactory for the farmers of the day because they made use of the land for grazing sheep on or for growing other and smaller fruit or vegetable crops between the trees – the practice of inter-cropping. But these huge standard trees ended up by having more vices than virtues and when mixed farms became more specialized and were changed over to grow only fruit, something had to be done to make fruit farming more profitable. The obvious step was to get smaller trees which could be planted closer together and which would carry heavier crops at an earlier date.

EXPERIMENTS IN STYLES

During the first quarter of the century these were developed and became available at the same time as newer and more precocious varieties were being raised, though Cox and Bramley were still the favourites. This was the time when bush trees came into the picture and soon standards and half-standards were almost forgotten. The bush trees could be planted 4 to 4.5 m (12 to 15 ft) apart to give 200 to 300 per acre.

With the bush trees came a new pruning method devised at East Malling, the renewal system. This was something of a compromise between the rather inexact regulated method and the time-consuming spur system (described previously on pages 52 to 54). The glory of the renewal pruning is that its severity can be altered to suit the size and vigour of the tree whereby weak trees are pruned somewhat harder than strong ones to get them to grow.

In essence, there is still a main framework of branches, as in the spur system, but this is furnished with temporary fruiting branches which are cut out when they have become too big or ceased to fruit well; hence 'renewal'. It was the first system that had been devised specifically to encourage early fruiting and the continuous production of high-quality apples.

The combination of renewal pruning and bush trees was fine where there was plenty of land but, gradually, the average size of fruit farms became smaller and smaller until by the 1950s there were people making a living from less than 8 hectares (20 acres) of trees. Before long we were back where we had been 20 years previously. Smaller and more profitable trees were again being demanded as profit margins became tighter.

At this point, many of the harder pressed fruit growers turned the clock back and started to grow apples and pears in the form of cordons, palmettes and dwarf pyramids. These had been in common use in the large walled gardens of earlier times but had never been considered a commercial proposition because of the time and expertise required to look after them. However, the squeeze meant that previously uneconomical methods might have a future after all. The research stations had already been doing work on the different methods of pruning which these intensive growing methods needed, and soon came up with a greatly simplified and quicker version of Lorette's system. Summer pruning was still necessary but the improved crops more than made up for this.

Renewal-pruned tree

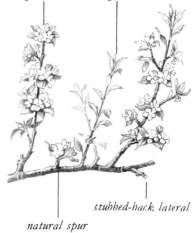

two-year lateral   *one-year lateral*

stubbed-back lateral

natural spur

Detail of renewal-pruned branch

Fifty years ago or so, it was common to have stock grazing between the rows of apple trees. This practice is still to be seen in some cider apple orchards particularly in Normandy, France

All this was going on at the height of the drift from the land when farm workers on all types of holdings were being enticed away by the glittering wages that industry was offering. From what had been reasonably labour-intensive businesses, agriculture and horticulture were moving more into the realms of industry, with farms having a small number of skilled workers in place of quite a large number of hard-working but unskilled labourers.

The effect of this trend on fruit farmers, particularly those growing tree fruits, was considerably greater than it was on agriculture. Farmers deal with annual crops and can change their tactics and routines relatively quickly, but fruit growers may have to wait for 20 years before an orchard becomes uneconomical or else face enormous losses if they have to grub it up in the prime of its life. It was imperative to find more profitable ways of growing apples. There was no question of them being abandoned but changes had to be made to fit in with the new ways of life.

RIGHT *Oblique cordon trained apple trees – suitable for growing in the smallest garden* OPPOSITE *A spindle tree. This style of training is more suitable for commercial orchards than for gardens (see page 80)*

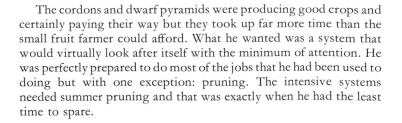

The cordons and dwarf pyramids were producing good crops and certainly paying their way but they took up far more time than the small fruit farmer could afford. What he wanted was a system that would virtually look after itself with the minimum of attention. He was perfectly prepared to do most of the jobs that he had been used to doing but with one exception: pruning. The intensive systems needed summer pruning and that was exactly when he had the least time to spare.

PILLAR TREES

In many respects Britain was behind the Continent because some of the possible solutions were already being developed in Holland and France. One idea in particular, though, was English – that of 'pillar trees'. This type of management required little skill and not a great deal of time but, above all, it was carried out in the winter. It consisted of a tree with a permanent central stem from which replaceable fruiting side shoots (laterals) were encouraged to grow. These were not pruned after their first year but were left to develop fruit buds along their entire length. At the end of their second year, the extension growth was cut off the end and, in the following summer, they carried fruit. In the winter they were cut out to make room for the younger shoots that were about to crop. This was a simple and straightforward solution with virtually no thinking involved at all.

Unfortunately, although all went well during the early life of the tree, it was a job to control its growth and things began to get out of hand, so the system fell out of fashion. It did, though, point the way towards a completely new system of management in commercial orchards, that of a specific tree form having its own system of pruning. The shape of the tree and the method of pruning went hand in hand. The tree had to be small and easy to manage; the pruning had to be simple and minimal.

THE SPINDLE SYSTEM

A method which possessed these criteria was the spindle system from Holland. This is still very popular and although it has undergone changes to cater for different circumstances, it is still essentially the same as it always was. The tree has a single central stem furnished with fruiting laterals but, instead of the relatively hard and intensive pruning needed by the pillar trees, virtually none is done at all.

The system makes use of the fact that a shoot will fruit sooner if it is left unpruned but that it will fruit even more quickly if it is tied down to a near-horizontal position. This makes a rather untidy tree in a garden but the system is, after all, not intended for domestic use; it is purely commercial.

The trees are planted and trained into rows by bending round and tying those shoots and branches that grow into the alleyway between rows. More or less the only pruning needed is the removal of the older branches when they have stopped fruiting at their best or when they have grown too high to manage.

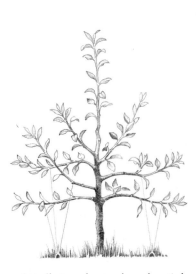

Spindle tree showing branches tied down to encourage horizontal growth

In fact, height is the most important factor of all in modern ways of growing apples. If the growth can be controlled, principally by the use of suitable rootstocks, and the height restricted by careful tying in and/or pruning, all ladderwork is avoided and with it an enormous amount of wasted time. This really is the key to present-day fruit farming. If spraying, picking and pruning can all be done easily and efficiently from the ground, the battle is half won already.

PYRAMIDS, CORDONS AND ESPALIERS

In domestic gardens, these smaller trees and intensive systems are having just as big an impact, but for a completely different reason. The ever-increasing problem in gardens is the lack of space but there is still a desire to grow fruit trees. Gardens are becoming smaller and smaller and the need for dwarf fruit trees is pressing. Pillars and spindles are not really what is called for, but dwarf pyramids, cordons and espaliers are admirably suited to garden conditions. Granted, they require a different approach and style of management to be at their best but, beyond this, they are no bother at all.

Cordons are normally planted and grown at an angle. This reduces the number of trees required for a given length of row but, more importantly, it also ensures that they crop sooner, as is shown with the tying down of spindles. Cordons are kept to a single stem and all laterals are pruned back in the summer. This keeps them restricted and fruiting well.

Dwarf pyramid

A spiral cordon tree. Despite its intricate appearance it is pruned in much the same way as a single cordon

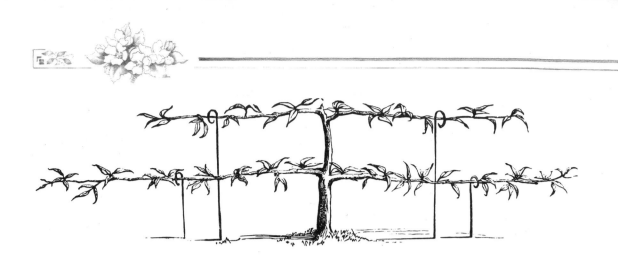

A two-tiered espalier. Espaliers can be trained against a wall or fence

Espaliers are slightly more complicated in that they have a central vertical stem from which are trained horizontal pairs of branches about 30 cm (1 ft) apart. Once the shape has been laid down and the branches started, pruning is essentially the same as for cordons.

Both these types of tree can be grown either in the open garden or against a sunny wall or fence but in both cases they need the permanent support of horizontal wires. When growing cordons, canes are tied to these wires and the tree is tied to the cane. With espaliers, the wires are spaced so that a pair of branches can be trained along each. Cordons and espaliers can be grown to any height but it is usual to keep them to about 2 m (6 to 7 ft) so that they are easy to manage.

You will remember that a new and very dwarfing rootstock called M 27 was mentioned earlier. This could well be the answer to many problems in the smaller gardens but care must be taken to ensure that the growing conditions, especially the soil, are good. If the tree has to struggle for its existence, results will be far from satisfactory.

Apple trees in pots

Raising apple trees and many other kinds of fruit in pots is a perfectly feasible, successful and attractive way of growing fruit in gardens. All that is really needed is a patio or somewhere similar to stand the pots during the summer and either a greenhouse or a conservatory to give protection when it is needed.

The main thing is to make sure that you purchase a tree on rootstock M 9 or M 27. These are really the only two rootstocks with characteristics that are dwarfing enough to make it possible. Although of lesser importance, it is not always wise to grow a vigorous variety like Bramley even on a dwarfing stock. It could prove too much for the average gardener to cope with. Family trees grown in pots are excellent, provided that they are on the right rootstock.

A point to remember is that trees in pots or other containers need a lot more looking after than if planted in the open ground. This may sound obvious now but it can easily be overlooked at the time when

Apple blossom is extremely decorative and tends to be under-rated – yet it adds much beauty to garden and countryside alike in spring

there are other things on your mind. It's a bit like having a dog: if you are not prepared to look after it, why get one in the first place? On the plus side, the obvious advantage of having your apple trees in pots is that they take up virtually no room at all. This has clear benefits in small gardens but, even in large ones, it enables you to grow far more trees and varieties.

Having mobile trees can be a great help at certain times of the year and this is really one of the main advantages of the system. The ability to move the trees under cover when the weather could damage them can make all the difference. The important period is in the spring during blossom time when a sharp frost can put paid to the fruit prospects for the year. If a frost is forecast, the trees can be brought under cover. Then again, when the fruit is nearing maturity, it may be helpful to move the trees either out of the wind or, in a bad year, into a warmer position. There is always the show-off value of being able to bring good-looking trees into the foreground or move tatty ones out of sight.

Apple trees perform well in containers if you follow the 'Points to Remember'

The pot that the tree is going to be grown in must have drainage holes and should never be too large to start off with. A 25-cm (10-in) pot is perfectly adequate. This can progress to a 30- or 35-cm (12- or 14-in) pot as the tree grows. Clay pots are normally better than plastic because of the extra weight and stability. They also have a better buffering action against sudden changes in temperature and are, on the whole, stronger. But if you prefer something lighter, go for a modern plastic container rather than a pot. These can be very serviceable and yet pleasant to look at. The most suitable kinds of compost are those based on loam. They do not have to be one of the John Innes types but must be of a reliable make. If using a John Innes potting compost, Nos. 2 or 3 are equally suitable. Loam-based composts are heavier, giving better stability, and tend to be more suitable for trees than the peat composts.

The best shape for an apple tree in a pot is a dwarf pyramid. It makes an attractive-looking tree as well as being an efficient one. Once the tree is established, repot every other year.

Feeding is possibly the most important routine job of all. With newly planted trees, the fertilizer in the compost will last for a good few weeks but in May/June, after the fruit has set, a dressing of a well-balanced feed, such as Growmore, should be given. If the leaves begin to turn pale in the summer, give it a regular liquid feed: the last thing that must be allowed to happen is that the tree should be starved when it is carrying a crop.

Trees in containers must be overwintered outdoors by plunging the pots in peat, bark or straw to stop the compost from freezing solid. Never try to coddle them by bringing them under cover; they don't appreciate it and may even start growing out of season.

That is really all there is to growing apples in pots except, of course, to make sure that the tree never runs short of water during the growing season.

Pollination and fertilization

Exactly when it was discovered that flowers had to be pollinated before fruit would form is uncertain but it must have been early in man's development. Many of the ancient races kept bees so it could not have escaped their notice that the bees visited the flowers. At some point two and two were put together and it was realized that pollination was taking place and that without it, fruit simply would not form. Ever since then, fruit growers have ensured that pollination is possible.

Pollination is simply the transference of pollen from the male parts (anthers) of a flower to the female part (stigma). For fruit to form, fertilization must follow. This is the process whereby the pollen grains germinate on the stigma, send a tube down into the ovary and fertilize it. It is therefore a two-part operation: first, pollination, then fertilization. If either fails to work, there will be no fruit. These two, shall we say, mechanical processes are certainly the most important but another factor is also necessary – that of compatibility.

In some cases, even if pollination takes place, nothing happens because the pollen fails to fertilize the ovary. There can be a number of reasons for this and whereas the mechanical side of the operation is fairly simple, the question of fertility is considerably less so.

For successful pollination there are two main conditions. The first is that there must be a sufficient number of pollinating insects to carry the pollen from flower to flower. In most cases this will be either bumble or honey bees. The other is that the right climatic conditions must exist for fertilization to take place after pollination. It really boils down to the need for sufficiently warm weather during the blossom period.

The main ways that fruit growers have for ensuring this is to keep their orchards well sheltered with effective windbreaks and to import colonies of bees to take care of the pollination. The fertility conditions are not quite so simple. Whilst some apple varieties will set a reasonable crop when fertilized by their own pollen, others will not. These properties are called self-fertility and self-infertility.

CROSS-FERTILIZATION

Although it may be interesting to know which group a particular variety belongs to, it is really of academic interest because all varieties crop best when fertilized by another variety (cross-fertilized). In gardens, this will normally happen automatically without any help. Bees travel a considerable distance when foraging and, except in the back of beyond, it is unlikely that they will fail to pick up pollen from a suitable neighbouring donor. In orchards, the situation may be rather different. There may be so many trees of one variety that adequate cross-pollination cannot occur. Growers overcame this difficulty by planting trees of other varieties amongst those of the main one simply to provide different pollen.

This works very well, on the whole, but it does introduce a few minor problems. The pickers have to remember that they are dealing with a different variety that has to be picked separately. It also means that the grower has trees in his orchard which might be unprofitable. There are several possible solutions but they all have to retain the pollinating variety.

One system that works quite well is to graft branches of a suitable pollinator on to a certain number of trees in the orchard. This does not supply as much pollen as a complete tree would but it is sufficient. However, it means that a tree contains a different variety and this is even harder for the pickers and pruners to manage.

With the coming of spindles, the trees were planted much closer together within the row but still with ample room between the rows. This meant that all the alleys ran in one direction so the obvious answer was to plant complete rows of the pollinator. Easy to manage, yes, but it still meant that trees were being grown that might not be wanted for their fruit.

But if unwanted trees were being grown, was it even certain that they were efficient pollinators? There was very little point in wasting space at the same time as growing something that failed to do a particularly good job. Would it not be preferable if there were fewer but better pollinators? At the same time, does a recognized variety of apple have to be used at all?

CRAB APPLES

Trials were carried out at the research stations, principally Long Ashton, with any apple that looked as though it would pollinate apple trees better than another variety. The answer came back loud and clear – certain crab apples worked like a charm. It would be too sweeping to say that all crab apples are good pollinators but some at least are excellent. The most successful are *Malus* 'Golden Hornet', *M. floribunda hillieri*, *M. aldenhamensis* and *M.* 'Winter Gold'.

This system has been used successfully for some years now but, at the moment, one of the most exciting lines of research is the use of synthetic hormones. These are sprayed on to the trees during the blossom period to cut pollination out of the sequence and simply fertilize the flowers chemically.

As can be imagined, there is still a long way to go. Not only must there be satisfactory results but there must also be a way of reducing the effectiveness of applied hormones so that only the required number of flowers are fertilized. Too many fruitlets are just as bad as too few. Still, these trials are certainly promising.

Growth regulators

The use of hormones for fertilizing the flowers to ensure good crops is a relatively new development in apple cultivation and is still in its infancy.

Not so is the use of growth-regulating hormones. These have been used since the mid-1960s in commercial orchards, probably arising as a by-product of weedkiller research. Once it had been discovered that applications of plant growth hormones could be used to upset the internal balance of unwanted plants such as weeds sufficiently to kill them, it was a relatively simple matter to develop similar materials for the *benefit* of plants. In the case of apples, it was found possible to reduce the rate of growth of a shoot without a consequent reduction in the number of leaves. This was of practical value, for it must be remembered that the number of leaves directly controls the number of flower buds.

It would be pointless simply to reduce the length of a shoot without any other benefit; a pair of secateurs would do that as efficiently. However, to reduce the length of a shoot and yet retain its fruiting capacity, and that of a tree, really can be regarded as a step forward. The main advantage obviously lies with vigorous varieties like Bramley but others also benefit, especially in a wet summer when growth is often at the expense of fruiting. Less pruning is another benefit, too. If the growth has been controlled chemically, there is less need to do it mechanically.

TOP *Double-worked Bramley's Seedling in a nursery row*
ABOVE *Compact columnar apple tree (see page 89)*

OPPOSITE Malus' *Golden Hornet'*

Something else that has developed over the years is the use of certain hormones to induce shoots to branch. It would appear from this that the growth of apple trees can be largely controlled by the timely use of hormone sprays. Fortunately this is not so; if it were, there would be very little enjoyment or skill involved in growing fruit and it would become more of a factory operation than horticulture. There are, however, distinct advantages to the commercial grower in being able to regulate the growth of his trees if the weather, or other factors, are not on his side. At the moment, growth regulators are unlikely to do away with the need for skill and knowledge but they can often take the business of earning a living out of the hands of Lady Luck. Incidentally, it must be said that although similar dwarfing chemicals are available to amateurs for use on certain pot plants, applying them to fruit trees will usually end in disaster.

There remains a lot of research to be done to improve things still further. We will never be able to succeed without having Nature on our side but we can at least try to shorten the odds so that the effect of disasters may be lessened or averted, even if never entirely eliminated.

Future prospects

There is no doubt that the irradiation of existing varieties is one of the most exciting and, one hopes, rewarding branches of current research. It might lead, for example, to a weaker growing Bramley that could be raised without difficulty in even the smallest of gardens, as well as having huge benefits for the commercial fruit farmer. Then again, we might end up with a Cox that blossomed later, to avoid spring frosts, and also matured later so that its season of use was extended. Changes of this sort may not be spectacular at first sight but they would make all the difference and are well within the realms of possibility.

PICKING BY MACHINE

From the commercial angle, one important line of research is in the growing of apples trees that can be picked by machines. It is another aspect that may sound rather futuristic but it has been carried out for many years on blackcurrants and raspberries. Although one of the problems is the likelihood of bruising the fruit, this is not insuperable if the trees are sufficiently small and uniform. They could then be picked without the fruit dropping from a height and being damaged.

This is not just a pipe-dream because, in the mid-1970s, the idea led to what were christened 'meadow orchards'. These consisted of trees no more than a metre (yard) high, planted very close together and cropping on a biennial basis.

The single shoot grew from the rootstock in the usual way in Year 1. This was treated with a suitable dwarfing agent to induce it to carry fruit in Year 2. All the fruiting shoots in the meadow orchard would be cut off, complete with their apples, by machine. The apples would go one way and the rubbish the other.

Year 3 would be the same as Year 1, except that more than one shoot would appear from each mini-tree. Thus, a biennial system would be established with the shoots growing one year and fruiting in the next, after which they would be cut down, rather like raspberries. It seemed a marvellous idea since picking and pruning would be carried out by machine in one operation and all irrigation and sprays would be applied by a system of permanent pipework and sprinklers. If need be, these could also be used for frost protection. On paper, there was hardly a fault.

However, even using the cheapest method of propagation available, just imagine what it would cost to buy and plant apple trees as close as 30 cm (12 in) apart in the row with 45 cm (18 in) between rows. At that spacing, we are talking in terms of around 29,000 trees per acre. Add to that the fact that each tree would only be cropping in alternate years. . . . Good as the idea appeared, trials proved it to be commercially impracticable.

In spite of these obvious objections, it is the sort of thing that might bob up again if a really cheap method of propagating apples can be discovered, for example by rooted cuttings or even seed.

COLUMNAR TREES

Something which is neither fanciful nor impossible is a trial that is currently going on at East Malling and elsewhere using what are called 'compact columnar trees'.

Some years ago, an apple tree of the Canadian variety MacIntosh threw a branch sport that was very straight and completely devoid of side shoots. Luckily, it was seen to be something quite out of the ordinary and so was propagated and given the name of Wijcik (*wich-ic*) after the farmer who discovered it. Clearly it had terrific possibilities. Unfortunately, Wijcik is a poor apple by U.K. standards so a breeding programme was set up in Britain to see if something could be done to improve the fruit whilst retaining the unique habit of growth. Cox was the obvious partner and, as luck would have it, about half the seedlings produced by the cross have the same columnar habit as Wijcik. Now all that remains is to find a seedling with an acceptable fruit. No easy task, but things are coming along well and several offspring are on trial.

The possibilities are endless. On the commercial side, trees of this sort would enable growers to plant pollinators that took up virtually no room at all. As fruiting varieties, they would need almost no pruning and could be suitable for mechanical harvesting. They would be planted no more than a 60 cm (2 ft) apart and would start cropping eighteen months later. These points would also make the trees absolutely ideal for gardeners, perhaps even more so than for growers.

The prospects for decorative and fruiting crab apples look especially good, too.

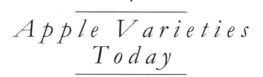

V

Apple Varieties Today

he history and development of apple varieties have been dominated by Cox's Orange Pippin and Bramley's Seedling. Nothing has really changed and, though other excellent new apples have come along to replace inferior and older varieties, these two still make up the greater part of the commercial acreage. In spite of their drawbacks, they are the two varieties that the public know, recognize, like and buy. What grower can ignore a guaranteed sale as strong as that?

Suitable for home growing

It is one thing to go into a shop and ask for a particular variety but quite another to grow it in your own garden. A different set of criteria applies and, as we have seen, Cox and Bramley have problems.

When considering what varieties to plant at home, the first consideration is bound to be whether you want a cooker or an eater. This is closely followed by the taste and characteristics of the fruit. Its appearance does not matter so much, but the season of use is very important. The choice rests between an early, mid-season or late variety. The first two are simple enough as the fruit is normally eaten straight off the tree, or very soon after picking. Late varieties, however, need to be stored and this raises the question of suitable conditions.

Then there are the characteristics of the tree. It is very little use buying a vigorous variety growing on a vigorous rootstock if you only have room for a small tree. Lack of space does not always

OPPOSITE
Fresh Picked Apples *by Alphonse Mucha (1860 1939)*

preclude a strong variety; for example, Bramley on M 9 or even M 27 rootstock can fit into most gardens but you may have difficulty in finding such a tree in your local garden centre. Still, do not let that deter you; there are plenty of specialist fruit nurseries. Failing that, there are, of course, several other varieties that have none of the faults of Bramley and are of almost as high a quality – Lane's Prince Albert, for example.

The amount of pruning that a tree needs may be important to you, as well as its suitability for growing in one of the trained forms for intensive cropping (cordon, espalier, or dwarf pyramid). One often thinks that there is not enough room for an apple in the garden and fails to realize that these forms take up no room at all. They can even be grown as a type of internal hedge in the garden to divide one section off from another.

The variety's flowering time may also matter; it it flowers early in the season it is likely to be hit by frost in a cold spring. It would be better to buy one less susceptible (see page 96). Also, problems may arise if a certain variety is prone to any particular pests or diseases. These are the sort of things to ask yourself when considering what to plant.

With this in mind, it always seems more helpful to have a list of the characteristics required followed by the varieties that meet them rather than having to wade through the varieties in search of their virtues and vices.

When dealing with Nature, the first lesson to learn is that the only things that fit neatly into pigeon holes are pigeons, whatever the statisticians would like us to believe. This may explain why, in the lists that follow, some varieties will appear in more than one single category. For instance, Discovery is listed as both a summer and an autumn apple because it ripens during August and September. Similarly, there are some varieties (Forge and Blenheim Orange, for example) which are dual-purpose, being suitable for both cooking and eating raw.

Not all the varieties mentioned will be generally available but they have all come from one nursery list or another, so are certainly obtainable at the time of writing. However a book specifically about apples would not be doing its job if the lesser-known ones were left out, though very obscure or unobtainable varieties have been omitted.

Those described as winter varieties will keep until the time shown by using the kind of storage facilities that are possible in most homes. In general, they should be kept above freezing point but as near to it as possible. The stated end of the season does not necessarily tally with the time when the fruit stops being available in the shops; commercial growers will have kept them in controlled atmosphere and temperature stores.

The question sometimes arises as to which are the best varieties for growing by one of the intensive methods. Most are perfectly satisfactory but it is normally best to avoid the vigorous ones unless on a dwarfing rootstock. Obviously the naturally small varieties are ideal.

Those varieties marked with a (C) on page 96 are cookers.

Double 'U' cordon

SUMMER (JULY/AUGUST)

EATERS

Beauty of Bath	Kerry Pippin
Devonshire Quarrenden	Merton Knave
Discovery	Miller's Seedling
George Cave	Worcester Pearmain
Irish Peach	

COOKERS

Arthur Turner	Grenadier
Early Victoria	

AUTUMN (SEPTEMBER — NOVEMBER)

EATERS

Allington Pippin	Katy (new)
Blenheim Orange	Kent (new)
Charles Ross	Kidd's Orange Red
Chivers Delight	Lord Lambourne
Cox's Orange Pippin	Merton Charm (new)
Discovery	Merton Knave (new)
Egremont Russet	Orleans Reinette
Ellison's Orange	Ribston Pippin
Epicure	St Edmund's Pippin
Fortune (Laxton's)	Spartan (new)
Gloster 69 (new)	Sunset
Golden Delicious	Tydeman's Early
Greensleeves (new)	Worcester Pearmain
James Grieve	

COOKERS

Bountiful (new)	Lord Derby
Forge	Peasgood Nonsuch
Grenadier	Rev. W. Wilks
Gloria Mundi	Stirling Castle
Golden Noble	

WINTER (DECEMBER ONWARDS)

EATERS

Allington Pippin	Jupiter (new)
Ashmead's Kernel	Kent (new)
Blenheim Orange	Kidd's Orange Red
Chivers Delight	Laxton's Superb
Cornish Gilliflower	Orleans Reinette
Court Pendu Plat	Pixie (new)
Cox's Orange Pippin	Ribston Pippin
Crispin	Rosemary Russet
D'Arcy Spice	Spartan (new)
Egremont Russet	Sturmer Pippin
Forge	Sunset
Golden Delicious	Suntan (new)
Gloster 69 (new)	Tydeman's Late Orange
Idared	Wyken Pippin
Jonagold (new)	

PAGE 94 TOP LEFT *Beauty of Bath;* TOP RIGHT *Early Victoria;* CENTRE LEFT *Egremont Russet;* CENTRE RIGHT *Golden Delicious;* BOTTOM LEFT *Greensleeves;* BOTTOM RIGHT *Jonagold*

PAGE 95 TOP LEFT *Jupiter;* TOP RIGHT *Orleans Reinette;* CENTRE LEFT *Wyken Pippin;* CENTRE RIGHT *Howgate Wonder;* BOTTOM LEFT *Lane's Prince Albert;* BOTTOM RIGHT *Katy*

COOKERS

Annie Elizabeth	Harvey
Bramley's Seedling	Howgate Wonder
Bountiful (new)	Lane's Prince Albert
Crawley Beauty	Lord Derby
Edward VII	Newton Wonder

LESS SUSCEPTIBLE TO FROST

Court Pendu Plat	James Grieve
Crawley Beauty (C)	Laxton's Superb
Edward VII (C)	Lord Derby (C)
Ellison's Orange	Newton Wonder (C)
Epicure	Worcester Pearmain

MOST SUSCEPTIBLE TO FROST

Allington Pippin	Devonshire Quarrenden
Blenheim Orange	Golden Noble (C)
Bramley's Seedling (C)	Lane's Prince Albert (C)
Charles Ross	Peasgood Nonsuch (C)
Cox's Orange Pippin	

MOST SUSCEPTIBLE TO SCAB

Allington Pippin	Early Victoria (C)
Beauty of Bath	James Grieve
Blenheim Orange	Newton Wonder (C)
Bramley's Seedling (C)	Rev. W. Wilks (C)
Cox's Orange Pippin	Worcester Pearmain

USEFUL IN A COLD DISTRICT

Allington Pippin	James Grieve
Court Pendu Plat	Lord Lambourne
Early Victoria (C)	Ribston Pippin

VIGOROUS GROWERS

Arthur Turner	Katy
Blenheim Orange	Laxton's Superb
Bramley's Seedling (C)	Newton Wonder (C)
Crispin	Orleans Reinette
Ellison's Orange	Suntan
Gloster 69	Tydeman's Late Orange
Irish Peach	

LESS THAN AVERAGE VIGOUR

Bountiful (C)	Lane's Prince Albert (C)
Court Pendu Plat	Miller's Seedling
Fortune	Rev. W. Wilks (C)
Greensleeves	Wyken Pippin

New varieties

Always a difficult word to describe and justify, for present purposes 'new' refers to varieties that have appeared in the 1960s and later or which are new to the garden market.

Most of the varieties mentioned above are well known and descriptions will be found in any good book or article dealing with apple varieties. But a lot of breeding work has been going on, both here and abroad, since the last war. This has led to a large number of new varieties finding their way into the shops both as fruit and also as trees in garden centres and nurseries. It would be well worthwhile looking at these, which have been included on the list above, because all are worthy of a place in the garden for one reason or another. They are all dessert varieties except for the cooker Bountiful.

GLOSTER 69

Raised in Germany soon after the war. A type of Delicious, but better. A long, bright red, late-keeping apple. Juicy and sweet with a pleasant flavour.

GREENSLEEVES

Raised at East Malling in 1966 and released to the public in the early 1980s. A James Grieve × Golden Delicious cross. Green to yellow skin. Will stay on the tree well into October without spoiling. Juicy with mild, refreshing flavour. Far superior to Golden Delicious.

JONAGOLD

An American variety with Golden Delicious and Jonathan as its parents. Introduced in 1968. Infinitely better flavour than either parent; juicy, sweet and rich.

JUPITER

Raised at East Malling in 1966 from a Cox × Starking cross. Somewhat angular fruit which stores well until January. Cox-like in flavour but heavier-yielding and more reliable. Darker and more coloured than Cox. No use as a pollinating variety because it produces sterile pollen.

KATY

Raised in Sweden in 1947 from a James Grieve × Worcester Pearmain cross. Started life as Katja but was changed to Katy for the U.K. market. Mid-season. Highly coloured on a yellow ground when ripe. Good texture and very pleasant flavour. The fruit is often found in the shops now.

KENT

Bred by Tydeman at East Malling in 1949 by crossing Cox with Jonathan. Well coloured on a pale green-yellow ground. Keeps until February. Well shaped and attractive. Fairly juicy with good aromatic flavour.

MERTON CHARM

Raised at John Innes Institute from a McIntosh × Cox cross and released in 1962. In season during September and October. Liable to bruising so of limited commercial value. Not highly coloured but of first-rate flavour and texture. Juicy and aromatic. Probably best when eaten off the tree.

MERTON KNAVE

A John Innes Institute seedling of Laxton's Early Crimson × Epicure dating from 1948. Very attractive and highly coloured red apple. Rather coarse-textured but juicy and sweet with a pleasant if not notable flavour.

PIXIE

Raised at the National Fruit Trials in 1947 from what is thought to be a Cox or Sunset pip. A high-quality late apple that will keep until March. Very similar to Cox in appearance. Crisp and juicy with a good aromatic flavour.

SPARTAN

A Canadian variety raised as long ago as 1926 but has only recently made its mark in Britain. A late apple, in season from November until late February. A rather odd purple colour when ripening but paler when actually ripe. Crisp, fine-textured and juicy; very aromatic.

SUNSET

A Cox seedling raised privately in 1918 but only now being recognized as one of the best alternatives to Cox, though rather too small for commercial growers. In season from October to December. Very Cox-like in appearance and taste.

SUNTAN

A Cox × Court Pendu Plat seedling from East Malling. Raised in 1955. Will keep until February and is usually at its best after Christmas. Large and with Cox's appearance but not such a fine flavour. No good as a pollinator (triploid).

BOUNTIFUL

The only recently introduced cooking apple. Raised at East Malling from a Cox pip that was probably pollinated by a Lane's Prince Albert. Released in 1985. A compact tree about half the size of Bramley. Crops early in life. Resistant to mildew. Good-sized fruit, mainly green but light in colour on the sunny side. An excellent cooker where space is limited.

OPPOSITE *Old apple varieties illustrated in the* Pomona Brittanica *(1807); crab apples (top row), cider apples (centre) and eating apples (bottom row)*

VI

Apples to Drink

large part of the popularity of apples is due to them being available in liquid form as well as for fillings for pies and tarts. In liquid form, they come in three 'strengths'. The health-giving apple juice might be considered to come first on the list, simply because it is the nearest thing to liquid apples. Apple juice is very big business now and West Germany is responsible for producing most of what we buy.

The other two main liquid forms of apples are cider and calvados. Most people know what cider is but calvados, on the other hand, is less familiar and rather special. As anyone who has visited Normandy will probably know, one of the districts of that region is called Calvados and it is from there that this delightful apple liquor comes. Quite simply, it is distilled cider. It comes in many disguises: the safest is in bottles from a French supermarket but, if you're feeling adventurous, then it's worth buying some at the orchard gate. Look out for the sign *cidre et calvados*.

It must have taken a certain amount of skill, equipment and trial and error to create beer or mead (a sort of honey beer) and yet, just think how easy it is to make cider. It was probably discovered when a pile of gathered apples began to go off and 'cider' oozed out of the bottom of the heap. What a discovery!

From these humble beginnings a flourishing, although nearly always a local, industry developed in many parts of Western Europe, particularly in Normandy and Brittany, the Basque country, in parts of Britain, Germany and Ireland. It was from these areas that, especially over the past two centuries, emigrants took their knowledge to the countries of the New World. Thus establishing a cider-making tradition that still flourishes today in North and South America, Australia and a number of other temperate countries.

OPPOSITE *Cider apple varieties illustrated in the* Herefordshire Pomona *(1876)*

There are several theories about how cider came to Britain. One is that it was brought by the Phoenicians to the tin miners of Cornwall in pre-Roman times. However, the first mention of cider being a European drink comes in the reign of Charlemagne. It appears to have been restricted to France, particularly the Normandy region, in those early days but, was introduced or reintroduced into Britain soon after the Norman Conquest of 1066.

It is generally assumed that either the drink or the knowledge of making it came over with the monks that built Forde Abbey near Axminster soon after the invasion. They probably also brought over either trees or grafts of the varieties that they used in France. Cider was manufactured primarily for use within the Abbey but any surplus would have been sold off. So it was probably the monasteries that were responsible for its spread and popularity in the 11th and 12th centuries. Cider making appears to have caught on in a big way for by the middle of the 13th century, it seems that many manor houses as far north as Yorkshire had either cider orchards or the wherewithal to make cider. The better equipped had both; although there was no real cider industry as such. By Chaucer's time it was a well established drink.

However, by medieval times cider was being produced commercially in Kent; much of it being supplied to ships bound on long voyages. We are not told why this was but it would be nice to think of it as light relief for the crew, though it was more likely to have been a palliative for scurvy.

At the same time, it also became the habit of farmers to part-pay their workers with cider; a custom that was still practiced as recently as the 1920s. This wasn't as mean as it may sound because farm cider was a very wholesome drink and thirst quenching at the same time; something that was most necessary at harvest time.

During the 14th and 15th centuries, the cultivation of fruit, and therefore cider, went through a lean patch. The influence of the

RIGHT *Apples were taken to the mill by horse-drawn wagon (see page 108)*

Plate XXIX.

1. White Styre

2. Styre Wilding

3. Eggleton Styre.

4. Skyrme's Kernel.

5. Cider Lady's Finger.

6. Gennet Moyle

7. Bromley.

8. Red Royal

réyne. Chromolith. Brussels.

F Stackhouse Acton & Edith E.Bull.
for The Woolhope Club.

Normans was largely over and there was no real enthusiasm for fruit as a crop. However, this changed in the 16th century when Richard Harris started to take a much more responsible attitude about fruit growing (see page 32). His influence led to cider orchards being established in the West Country and also in Kent.

The increase in apple growing had an interesting effect on British agriculture. Whereas previously most of the barley produced went into the brewing industry, the popularity of cider and fall-off in beer meant that there was a lot more beer and barley available for export. So great was the rise in popularity of cider that more apple trees were planted in the middle of the 17th century than in all the previous several hundred years. This reached such a state that Gloucestershire, Herefordshire, Worcestershire, Sussex and Kent were all producing enough apples to supply their own needs all the year round; both for fruit and for cider. By the 1720s, Devon had come into the cider picture and took a big step forward by introducing smaller trees.

Cider apple varieties

Cider was originally made from varieties of crab apple but, by 1204, the English Pearmain had appeared and was still an important cider apple in 1657, 400 years later. Crab seedlings were also used as rootstocks because they gave rise to the longest lived trees; which were also the largest and slowest to bear. It was found that seedlings grown from actual varieties made smaller trees that bore crops sooner and it was these that were most probably used in Devon.

Perhaps the oldest named varieties were Genet Moyle and the Codlin but only the best producers used these. The normal drill was to use any variety available and this gave rise to some pretty rough cider. By the mid-1600s opinions were starting to firm up so that Pearmain was considered to be a good variety, Genet Moyle was better and one called Redstreak the best of all. Redstreak was introduced by Lord Scudamore who brought over a number of apple varieties from France (see page 36). Two other apples previously mentioned, Harvey and Golden Pippin, were also popular cider varieties. Perhaps the most famous variety of all, Foxwhelp, appeared in Hereford during the 17th century and is still grown today. In the late 18th century, the variety Kingston Black was found near Taunton and really came to prominence when it reached Hereford in about 1820. It is not widely grown now but is still considered excellent for vintage ciders.

The upsurge in popularity that cider apple growing underwent in the 17th century was on the wane by the mid-18th century in favour of arable and livestock farming. By the mid-19th century it was rather an 'also ran' and had slipped into the 'home-brew' status. This was thought at the time to be due in no small part to the rather puritanical writers of the day decrying the amount of drunkeness and debauchery that prevailed. However, it is far more likely that the sharp and lethal spring frosts every year from 1800–1817 played a much greater part. Eighteen years without a decent crop would have

ABOVE *A pair of English lead crystal cider glasses*
OPPOSITE *Building a cheese using 'hairs' of sacking (see page 108)*

been a severe blow to the cider industry, as well as to a farmer's finances. The end of the Napoleonic Wars added further to the decline of cider apple growing for it meant that wine and brandy were freely available again. Herefordshire was the worst hit, being the leading county for cider production.

By 1876, the situation there was so bad that the Woolhope Naturalists' Field Club organised a survey of Hereford to see just how bad things were and as a direct result of this, new trees were grown from material supplied from the Royal Horticultural Society's gardens at Chiswick, amongst them Foxwhelp. In 1884, under the auspices of the Woolhope Naturalists' Field Club, two new varieties which are still grown today, Medaille d'Or and Michelin, were introduced from France. It is interesting to note that the Rev. C.H. Bulmer was a member of the survey team; probably the best known of all names in the cider world today, as then.

At the turn of the century interest in cider production was again aroused, mainly through the work of the survey, and the National Fruit and Cider Institute was set up. In 1912, this body joined forces with Bristol University and became the Long Ashton Research Station; until recently one of the top two fruit research establishments in Britain but now, sadly, working on other crops.

Amongst other things, Long Ashton tackled the problem of varieties. This situation had got completely out of control with farmers growing whatever they felt like with most of the varieties being poor and unnamed seedlings. By the mid-1930s, Long Ashton had sorted out much of the chaos so that at least growers could get a definite recommendation as to what they should be growing and be able to purchase these varieties. A fruit tree trial carried out at that time showed up the worth of a relatively unknown variety called Yarlington Mill – a seedling that came from Somerset in the late 19th century and possibly the doyenne of cider apples. Now this brings us up to date with the story of cider apples but it is worth noting that the wheel has turned and cider is again a popular drink.

The way in which the apples are grown has also changed and cider orchards are hard to distinguish from dessert and cooking apple orchards. The trees are small and require a minimum of upkeep and the fruit is harvested mechanically. Also, it is becoming increasingly the job of the big cider companies to grow their own fruit. No longer do they trudge round the district buying this and that from farms; they are now growing what they want and have complete control over the operation from the orchard to the bottle.

The production of cider

Having looked at the way in which cider apples have been grown over the years, let's see how the production side developed. Cider was essentially a farm product, most of which was drunk locally and only the surplus sold. It was pretty rough stuff. No yeasts were used and it was allowed to finish naturally. This meant that it was left until fermentation was complete and then stored and drunk. In this state it was completely flat; unlike beer which retains a certain sparkle.

An old stone cider press

However, in those early days, cider had one distinct advantage over even water; it was completely safe to drink. Water wasn't, it could be full of all sorts of harmful organisms, but nothing could survive the acid conditions of cider. Oddly enough, perry, the pear equivalent of cider, wasn't as popular mainly because you couldn't drink as much and still stay upright!

Cider apples tend to be rich in tannin and sugars; the tannin gives rise to the different flavours and the sugar to the strength. They cannot be eaten raw; they're much too sharp. In the West Country, the belief is firmly held that you can't make good cider out of anything other than special cider varieties; other regions think differently, and seem to succeed. The quality of farm cider really depended on whether the farmer blended his apples carefully or just put in anything available. Unfortunately, most followed the latter path.

The normal way of growing the fruit was to have orchards of standard trees in which livestock were allowed to graze until the autumn when the fruit dropped. After gathering, the apples were allowed to 'mature' in piles for anything up to a month or more.

Not all farmers had the facilities to press their own apples so transportable cider presses such as the one illustrated here used to travel around the country. This photograph was taken in 1934

When it came to the pulping, only completely black apples were removed, broken and bruised ones were allowed to remain. However, it was soon found that bruised apples contained less sugar and, thus, made cider that was less alcoholic. The apples were crushed by a number of different methods, the most efficient being by horse-driven mill. However, all the systems had one thing in common, it was fatal to allow the apples or the 'must' (apples after crushing) to come into contact with iron. In fact, it really proved to be fatal in some cases.

Not all farmers had the equipment to crush the apples and employed freelance operators who travelled the county with portable crushers and pressers. Pressing was the most skilled part of cider making as it involved wrapping the must in 'hairs' of sacking or coconut fibre and then building them up until there were eight to twelve in the press. This was then called a 'cheese' which was then pressed to obtain juice. After pressing, the apple juice was put in casks where, in two to three days, it would start fermenting. Fermentation would go on for a week, a couple of months or even until the spring. It didn't, and still doesn't, seem to effect the quality of the cider at all.

Although no yeast was added to the casks to induce fermentation, some farmers put in pieces of meat. This certainly aided the process and it is now thought that the meat added protein to the brew; though this wasn't fully understood at the time. Stories that dead rats

THE · VIRGIN · AND THE · APPLE · TREE

There is a long-held tradition that when the Virgin Mary was heavy with child, she came upon an apple tree laden with fruit. As she could not reach up, the tree lowered its branches so she could pick the fruits with ease. This is why, or so the story goes, that many apple trees still have branches which hang down.

Perhaps this legend is rooted in Teutonic mythology where heaven was likened to a vale of apple trees tended by the goddess Idun. The apples were the fruits of perpetual youth and gave the gods their immortality.

were added are pure fantasy. There are, however, several instances of farm and domestic animals falling unknown into vats of fermenting cider and the accident not being discovered until the brew was bottled. An interesting, if revolting, fact about all these cases is that all that remained were the bones of the animals, everything else had been assimilated into the cider. (One farmer lost a pig and then eventually found its bones in the bottom of the vat. He reckoned this cider was the best he had ever tasted!)

This farm cider had a fairly short life once fermentation had stopped. It was usually left to mature for about three months but then had to be drunk because the quality started to fall off after about a year. This never seemed to be a problem though. The popular name and image for farm cider is 'scrumpy'. In fact, scrumpy is cider whose alcohol has turned into acetic acid (vinegar). No wonder it has made such a name for itself! The end of farm cider as an industry came between the World Wars and was entirely due to economic reasons. Its demise on one farm is summed up beautifully by a Herefordshire farmer. 'I paid 'em to make it, and I paid 'em to drink it, and still the blighters weren't satisfied so I stopped making it.'

VII

Apple Recipes

It cannot be certain when apples were first used as an ingredient of cooked dishes but it must go back into the mists of history, and the use of this fruit in a fermented form or as a distilled spirit must be as old as the invention of wine. Certainly cider formed part of the basic diet of the British people in early times and apples have been included in some of the earliest recorded recipes.

A good dessert apple has long been a favoured food and gardeners in the past have often resorted to magic and superstition in order to obtain a sweeter fruit. Sweet-smelling flowers grown near to apple trees were meant to encourage sweet-tasting fruits. Red apples were supposed to be encouraged if red roses were planted alongside a tree. In the 15th century, holes were frequently bored into tree trunks and a coloured liquid poured in. Red fruits would be produced, it was hoped, from an injection of red liquid and golden apples from an injection of yellow.

Another long-held superstition was that if the apples weren't 'Christened' with rain on 15th July, St Swithen's day, the harvest would be worthless.

Today apples, green, red or yellow, are a staple ingredient of many of our most delicious dishes; time has done nothing to diminish their popular lead over other cooked fruits.

Here are a selection of some traditional recipes in which apples are an indispensable ingredient.

OPPOSITE
Preparing for Dinner *by Edward Davis (1833–67)*

PORK · IN · CIDER

Serves 4

Cider is used in cooking in traditional dishes from the west of England as well as from Normandy and from the apple-growing states of America. Savoury casseroles prepared with cider instead of wine are very good, and rich ingredients, such as pork or mackerel, benefit from the zest which it contributes to the finished dish. In fact, pigs reared in the West Country were fattened on fallen apples and on many farms the pigs were allowed to roam the orchards. This fruity diet produced a joint of pork which offered a unique and superior flavour.

Although pork chops are the more common and traditional cut for cooking in cider, this hearty stew of meat and vegetables is just as successful.

METRIC · IMPERIAL · AMERICAN

1 large onion, sliced

2 tablespoons/2 tablespoons/3 tablespoons oil

675 g/1½ lb/1½ lb lean pork, cubed

25 g/1 oz/¼ cup flour

salt and pepper

600 ml/1 pint/2½ cups dry cider

a few sprigs of rosemary

225 g/8 oz/½ lb carrots, sliced

100 g /4 oz/1 cup button mushrooms, thickly sliced

450 g/1 lb/1 lb small new potatoes

bay leaf

Fry the onion in the oil in a flameproof casserole until soft but not browned. Add the cubes of pork and brown them all over. Stir in the flour, add plenty of seasoning and cook for a minute before pouring in the cider. Bring to the boil, stirring all the time, then add the rosemary and the vegetables. Reduce the heat and cover the pan. Leave to simmer gently for 45–50 minutes. Check that the pork is cooked, taste and adjust the seasoning before serving with roast potatoes, baked potatoes or some buttered pasta.

VARIATION

If you like, omit the potatoes and carrots; instead add a few raisins and cook as above. Peel, core and thickly slice 2 dessert apples then add them to the casserole about 10 minutes before the end of the cooking time. Serve with brown rice and peas.

ROAST · PHEASANT · FLAMBÉED WITH · CALVADOS

— ❧ —

Serves 4

Calvados is an apple brandy from Normandy where it is used in a variety of traditional dishes, including recipes for pork and fish as well as game. This is a delicious and very special way of serving pheasant, ideal for a dinner party.

METRIC·IMPERIAL·AMERICAN

2 pheasant, preferably hen birds, plucked and cleaned ready for the oven

6 rashers (slices) fatty streaky bacon

1 large onion, quartered

salt and pepper

50 g/2 oz/$\frac{1}{4}$ cup butter

4 tart dessert apples, peeled, cored and thickly sliced

2 tablespoons/2 tablespoons/3 tablespoons flour

150 ml/$\frac{1}{4}$ pint/$\frac{2}{3}$ cup red wine

150 ml/$\frac{1}{4}$ pint/$\frac{2}{3}$ cup single (light) cream

4 tablespoons/4 tablespoons/$\frac{1}{3}$ cup Calvados

2 tablespoons/2 tablespoons/3 tablespoons chopped parsley

Place the birds in a roasting tin and top each with 3 rashers (slices) bacon, laying them over the breast to prevent the meat from drying out during cooking. Tuck the onion quarters around the birds and sprinkle with seasoning. Dot with the butter and roast in a moderately hot oven (190 C, 375 F, gas 5) for about 1$\frac{1}{2}$ hours or until the birds are cooked through and tender. Add the apples to the tin 15–20 minutes before the end of the cooking time.

Have ready a warmed serving dish. Transfer the pheasants to the dish and remove the bacon. Place the apples in a separate dish. Keep hot while making the sauce.

Stir the flour into the pan juices and cook over a moderate heat for a few minutes. Stir in the wine and bring to the boil, stirring all the time. Remove from the heat, stir in the cream and adjust the seasoning, then reheat gently without boiling. Pour the sauce around the pheasants and arrange the apples in the dish.

Heat the Calvados so that it is just warm, set it alight and pour it over the pheasants, then serve while flaming. Make game chips (very thinly sliced and deep-fried potatoes) to complement the roast if you like.

HEAVEN · AND · EARTH

———⋘✦⋙———

Serves 4-6

The name of this dish is translated directly from the German title
Himmel und Erde. *Tart apples are cooked with potatoes in a typical
German vegetable dish. Onions, and sometimes bacon, are added and
traditionally the potatoes and apples are mashed together. The authentic
accompaniment would be fried German blood sausage but any good sausage
will go well – why not try serving fried slices of black pudding with the
Heaven and Earth?*

METRIC·IMPERIAL·AMERICAN
1 kg/2 lb/2 lb potatoes
salt and pepper
450 g/1 lb/1 lb cooking apples, peeled and cored
100 g/4 oz/6 slices fatty smoked bacon, rind removed and chopped
a little chopped parsley

Cut the potatoes into large cubes and cook them in boiling salted water
until tender – about 15 minutes. Drain thoroughly.

Cut the apples into chunks. In a heavy-based frying pan, heat the
bacon over a fairly low heat until the fat runs, then continue to cook
until the bacon is crisp. Use a slotted spoon to remove the bacon from
the pan, drain on absorbent kitchen paper and reserve.

Add the apples and potatoes to the fat remaining in the pan and
cook, turning occasionally, until the apples are tender and the potatoes
are hot and slightly browned. Sprinkle with seasoning to taste and the
parsley, then sprinkle the crisp bacon over and serve at once.

AMERICAN · APPLE · SAUCE CAKE

— ❧ —

Makes 1 20-cm/8-in cake

For this recipe you will need apples which fall to a fine purée or sauce when they are cooked because it is this smooth sauce which moistens and flavours the cake. If you cannot obtain pecan nuts, then use walnuts instead.

METRIC·IMPERIAL·AMERICAN

1 kg/$1\frac{1}{4}$ lb/$1\frac{1}{4}$ lb cooking apples, peeled, cored and chopped

2 tablespoons/2 tablespoons/3 tablespoons water

100 g/4 oz/$\frac{1}{2}$ cup butter or margarine

175 g/6 oz/$\frac{3}{4}$ cup sugar

1 teaspoon vanilla essence

1 egg

450 g/1 lb/4 cups plain flour

$\frac{1}{2}$ teaspoon ground cinnamon

1 teaspoon ground mixed spice

$\frac{1}{2}$ teaspoon ground cloves

$1\frac{1}{2}$ teaspoon bicarbonate of soda (baking soda)

225 g/8 oz/$1\frac{1}{4}$ cups raisins

225 g/8 oz/$1\frac{1}{4}$ cups pecan nuts, roughly chopped

Cook the apples with the water until they are reduced to a purée (sauce), then beat well until smooth and leave to cool completely.

Beat the butter or margarine with the sugar until pale and soft, then beat in the vanilla essence, cooled apple purée and egg. Sift the flour, spices and bicarbonate of soda. Stir in the raisins and nuts, then stir these dry ingredients into the mixture. Spoon into a well-greased loose-bottomed 20-cm/8-in cake tin. Level the surface and bake in a moderate oven (180 C, 350 F, gas 4) for $1\frac{1}{2}$–$1\frac{3}{4}$ hours. Check the cake to make sure that it does not become too brown on top after $1\frac{1}{4}$ hours. If it is dark, then cover it loosely with a piece of cooking foil. When cooked, a metal skewer inserted into the middle of the cake should come out clean of mixture.

Leave in the tin for a couple of minutes, then turn the cake out on to a wire rack to cool completely. If you like, the cake can be served warm with cream and extra apple sauce.

OPPOSITE *Scrumping for apples*

APPLE · STRUDEL

Serves 6-8

The preparation of the fine strudel pastry is an art which is best left to the professionals or those well experienced in the technique. Instead, philo pastry can be used – this is available from delicatessens where it is often sold frozen.

METRIC·IMPERIAL·AMERICAN

2 large sheets philo pastry

50 g/2 oz/¼ cup butter

450 g/1 lb/1 lb cooking apples, peeled, cored and finely chopped

50 g/2 oz/¼ cup raisins

50 g/2 oz/¼ cup sugar

grated rind of 1 lemon

50 g/2 oz/¼ cup fine dry breadcrumbs

icing sugar to dust

Lay one sheet of pastry flat on the work surface, brush it lightly with butter, than lay the second sheet on top. Mix the apples, raisins, sugar and lemon rind. Brush the second sheet of pastry with butter and sprinkle the breadcrumbs evenly over it. Spread the apple mixture over the top leaving a border all round the edge. Then roll the pastry up from the short end to enclose the apple filling completely as for a Swiss roll. Lift the strudel on to a greased baking tray and press the ends together to seal in the filling. Brush the top with the remaining butter and bake in a hot oven (220 C, 450 F, gas 7) for 15 minutes. Reduce the temperature to moderately hot (190 C, 375 F, gas 5) and cook for a further 20–30 minutes. Remove from the oven and dust immediately with icing sugar. Best served warm.

OPPOSITE
A was an apple-pie,
B bit it,
C cut it,
D dealt it,
E eat it,
F fought for it,
G got it, . . .
Part of a 17th-century nursery
rhyme. The title page from Kate
Greenaway's book (1886)

D A N I S H · A P P L E · C A K E

Serves 6-8

Apple purée or sauce is used extensively in Denmark to make desserts as well as in baking. One of the nicest fillings for Danish pastries is an apple purée. Serve this unusual 'cake' hot or cold, with plenty of whipped cream.

METRIC · IMPERIAL · AMERICAN

1 kg/2 lb/2 lb cooking apples, peeled, cored and chopped

2 tablespoons/2 tablespoons/3 tablespoons water

75 g/3 oz/6 tablespoons sugar

75 g/3 oz/6 tablespoons butter, melted

grated rind of 1 lemon

225 g/8 oz/1 cup dry, very lightly browned breadcrumbs

50 g/2 oz/$\frac{1}{4}$ cup demerara sugar

Cook the apples with the water, sugar and lemon rind until they are reduced to a purée (sauce). Boil for a few minutes to thicken the mixture, then remove from the heat and beat thoroughly.

Grease and base-line a deep 20-cm/8-in cake tin. Use some of the butter to grease the paper generously. Mix the breadcrumbs with the sugar, then layer with the apple purée in the tin, beginning and ending with crumbs. Pour the remaining butter over the top and bake in a moderately hot oven (190 C, 375 F, gas 5) for about 30 minutes, or until the top is golden. Turn out to serve hot, or leave in the tin until warm, then turn out to cool completely and serve cold.

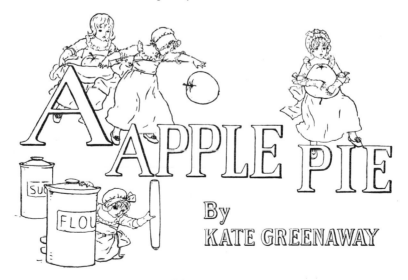

TARTE · TATIN

Serves 6

The tradition of classifying apples as 'cooking' or 'eating' is a British one and recipes taken from other countries often rely upon the firmness of dessert apples for success. This French apple tart is a good example – it is essential that the apples do not fall during cooking so that the tart will reveal a caramel-coated filling when inverted on to the serving dish.

METRIC·IMPERIAL·AMERICAN

Pastry

175 g/6 oz/1½ cups plain flour

75 g/3 oz/6 tablespoons butter

25 g/1 oz/2 tablespoons sugar

1 egg yolk

Filling

1 kg/2 lb/2 lb dessert apples

50 g/2 oz/¼ cup butter

50 g/2 oz/¼ cup sugar

Sift the flour into a bowl, then rub in the butter until the mixture resembles fine breadcrumbs. Stir in the sugar, then work in the egg yolk to make a short, sweet pastry. Chill briefly.

Peel, core and quarter the apples. Melt the butter in an 18-cm/7-in ovenproof pan – a copper pan or deep sandwich tin will do if you do not have a special tarte tatin pan. Stir the sugar into the butter and cook gently until it dissolves and begins to caramelize. Remove from the heat, then pack the pieces of apple tightly into the pan.

Roll out the pastry into a circle slightly larger than the top of the pan. Lift it over the apples and tuck the edges down into the pan. Bake in a moderately hot oven (200 C, 400 F, gas 6) for about 30 minutes, or until the pastry is cooked. Invert the Tarte tatin on to a serving plate and serve hot or warm with cream.

OPPOSITE *The children of the Cobham Family from a painting by Hans Eworth (1540–73)*

ENGLISH · APPLE · LOAF CAKE

Makes 1 small loaf

This is a type of tea bread, made with grated apples and cider. It is delicious served warm with butter and clotted cream or it will keep well in an airtight container for up to a week.

METRIC · IMPERIAL · AMERICAN
350 g/12 oz/3 cups plain flour
3 teaspoons baking powder
100 g/4 oz/$\frac{1}{2}$ cup butter or margarine
100 g/4 oz/$\frac{1}{2}$ cup sugar
1 teaspoon ground mixed spice
$\frac{1}{2}$ teaspoon ground cloves
175 g/6 oz/scant $\frac{1}{2}$ lb cooking apples (1 medium sized), peeled, cored and grated
50 ml/2 fl oz/$\frac{1}{4}$ cup medium-sweet cider

Sift the flour into a bowl with the baking powder. Rub in the butter or margarine until the mixture resembles fine breadcrumbs. Stir in the sugar and spices, then add the apples and mix well. Stir in the cider to make a stiff mixture.

Turn the mixture into a well-greased 1-kg/2-lb loaf tin and bake in a moderately hot oven (190 C, 375 F, gas 5) for about 1 hour, or until risen, golden on top and firm to the touch. When cooked, a skewer inserted into the middle of the loaf should come out clean. Leave in the tin for a few minutes, then turn out to cool on a wire rack. Serve cut into thick slices and buttered. Good warm or cold.

TOFFEE · APPLES

Makes about 10

For bonfire night toffee apples are a must! These are very easy to prepare, using a simple caramel coating which is crunchy and light. It is best to make the apples on the same day as they are to be eaten, so that they do not have time to soften slightly and become sticky.

METRIC·IMPERIAL·AMERICAN
450 g / 1 lb / 2 cups granulated sugar
600 ml / 1 pint / 2½ cups water
about 10 dessert apples
wooden sticks

Put the sugar in a saucepan and pour in the water. Stir over a low heat until the sugar dissolves, then bring to the boil and boil until the syrup changes colour and a light caramel is formed. The mixture will continue to cook after the heat is turned off, so don't let it turn too dark.

While the caramel is cooking, put the apples on sticks and prepare a sheet of oiled greaseproof paper on a baking tray. As soon as the caramel is ready dip the apples in it, twirling them round to coat them evenly. Dip them a second time to give a thick coating, then put them on the oiled paper to cool completely.

APPLE · BUTTER

Makes about 1.5 kg/3 lb

Fruit butters and cheeses were great favourites in Victorian times, served with thin bread and butter for afternoon tea. Generally, the yield of this type of preserve is small for the quantities of fruit and sugar which are required but the result is a sweet spread which can be used sparingly. As well as using this butter instead of jam, try using it to flavour desserts – simple creams or thick natural yogurt.
Fallen apples are ideal for this recipe and a mixture of tart and sweet apples can be used. Some crab apples can also be added. If the fruit requires very lengthy cooking you may have to add extra water.

METRIC·IMPERIAL·AMERICAN
1.25 kg/2⅓ lb/2½ lb apples, chopped
1 cinnamon stick
1 teaspoon grated nutmeg
1 lemon, chopped
600 ml/1 pint/2½ cups water
about 575 g/1¼ lb/2½ cups sugar

There is no need to peel the apples. Put them in a large pan with the spices and lemon, then pour in the water. Bring to the boil, reduce the heat and simmer for about 1 hour, or until the fruit is reduced to a pulp.

Press through a fine sieve and weigh the resulting pulp. For each 450 g/1 lb purée add 350 g/12 oz sugar. Stir this mixture over a low heat until the sugar dissolves, then bring to the boil and boil steadily for about 30 minutes. The mixture should be reduced to about half its original volume and it should be thick and creamy. Stir frequently during cooking to prevent the apple from burning.

Spoon into hot jars and cover with waxed discs, waxed sides down. When cool, top with lids and label. The preserve will keep for three to four months.

OPPOSITE *Victorian apple seller*
PAGE 122 *Detail of an advertisement for Bulmer's cider*
PAGE 123 *Mixed apples*
PAGE 124 *Marquetry panel from a desk (c.1779) made in France*

MULLED · CIDER

— ❦ —

Serves 6-8

This is quite heart-warming and it is an inexpensive alternative to mulled wine. Select a dry or medium-sweet cider according to taste.

METRIC·IMPERIAL·AMERICAN
1 large orange
10–12 cloves
1 cinnamon stick
1 bottle cider
150 ml/$\frac{1}{4}$ pint/$\frac{2}{3}$ cup rum (optional)

Stud the orange with the cloves, then place it in a large pan. Add the cinnamon stick and pour in the cider. Stir in the rum, if used, and cover the pan. Heat over the lowest possible setting for about 45–60 minutes, or longer. Do not allow the cider to become too hot. For parties or larger gatherings of people, top up the cider, adding extra rum and keep the pot warming all the while. Serve in heatproof glasses.

CRAB · APPLE · JELLY

Often found growing wild in the country hedgerows, small ruddy crab apples make a mouth watering jelly. Serve it as an accompaniment to savoury dishes – roast pork, gammon, with lamb or pork chops and cold cooked ham – or have some on hot buttered toast for breakfast or tea.

METRIC·IMPERIAL·AMERICAN
1.75 kg/4 lb/4 lb crab apples, rinsed and roughly chopped
1.15 litres/2 pints/5 cups water
juice of 2 lemons
4 cloves
sugar

Put the apples in a large saucepan with the water and the lemon juice. Bring to the boil, then reduce the heat and cook the fruit for about 1½ hours, or until pulpy.

Leave the fruit to cool slightly, then strain it through a jelly bag overnight. Do not squeeze the bag next day or you will have a cloudy jelly. Measure the resulting liquid, pour it into a saucepan and add 450 g/1 lb/2 cups sugar (granulated or preserving) for each 600 ml/ 1 pint/2½ cups. Stir over a low heat until the sugar dissolves, then bring to the boil and boil hard until setting point is reached.

To test for setting, drop a little of the jelly on a very cold saucer: it should form a skin which wrinkles when pushed after a couple of minutes. Alternatively use a sugar thermometer which will read 104 C/220 F when setting point is reached.

Pour the jelly into warmed, clean pots and cover with waxed discs, putting the waxed sides down. Leave until cold before covering with lids or cellophane tops. The yield will depend on the juice from the fruit. The jelly will keep for three to four months.

APPLE · CHUTNEY

— ❧ —

Makes about 2.25 kg / 5 lb / 5 lb

The ideas for the savoury chutneys, which are traditionally prepared every autumn to preserve the fruits of the season, come from the Indian condiments known as chatni. *These highly spiced side dishes were adapted to suit the western palate and the ingredients which were commonly available. This recipe is for a simple, fairly sweet chutney, so reduce the quantity of sugar by about $50\,g/2\,oz/\frac{1}{4}$ cup if you favour a tart preserve.*

METRIC · IMPERIAL · AMERICAN

1.5 kg / 3 lb / 3 lb cooking apples, peeled, cored and chopped

450 g / 1 lb / 1 lb onions, chopped

$50\,g/2\,oz/\frac{1}{4}$ cup raisins

$50\,g/2\,oz/\frac{1}{4}$ cup fresh root ginger, grated

1 green pepper, deseeded and chopped

1 tablespoon mustard powder

1 tablespoon ground coriander

3 cloves garlic, crushed

$275\,g/10\,oz/1\frac{1}{4}$ cups demerara sugar

$600\,ml/1$ pint $/2\frac{1}{2}$ cups vinegar

Put all the ingredients together in a large saucepan and mix well. Bring slowly to the boil, stirring occasionally, then reduce the heat and cover the pan. Leave the chutney to simmer for about 1 hour. Stir the mixture occasionally to prevent it sticking to the pan.

Have ready clean, warmed pots. Spoon the chutney into the pots and cover with waxed discs, waxed sides down. Top immediately with airtight lids. Label with the name and date. Store the chutney for a few weeks before sampling it, then it can be kept for up to six months.

Bibliography

BAKER, MARGARET, *The Gardener's Folklore*, David and Charles, London, 1977.

BUNYARD, EDWARD. A. *Handbook of Fruits (Apples and Pears)*, John Murray, London. 1920.

BULTITUDE, JOHN, *Apples: A Guide to the Identification of International Varieties*, MacMillan, London. 1983.

CHAMPAGNAT, PAUL, *The Pruning of Fruit Trees*, Crosby Lockwood, London. 1954.

GARNER, ROBERT JOHN. *The Grafter's Handbook*, Faber and Faber, London. 1967. 3rd ed. rev. and enl.

HOGG, ROBERT. *The Fruit Manual*, Journal of Horticulture Office. 1884. 5th ed.

HUXLEY, ANTHONY JULIAN. *An Illustrated History of Gardening*, Paddington Press for The Royal Horticultural Society, London. 1978.

LORETTE, LOUIS. *The Lorette System of Pruning*, John Lane the Bodley Head, London. 4th rev. ed. 1948.

LARGE, E.C. *The Advance of the Fungi*, Jonathan Cape, London. 1940.

New Larousse Encyclopedia of Mythology, Hamlyn. 1968. 2nd ed.

OPIE, I. AND P. (eds). *The Oxford Dictionary of Nursery Rhymes*, Oxford University Press, Oxford. 1951.

ROACH, F.A. *Cultivated Fruits of Britain: Their Origin and History*, Basil Blackwell, Oxford. 1985.

SMITH, MURIEL. *National Apple Register of the UK*, Ministry of Agriculture, Fisheries and Food, London. 1971.

TAYLOR, H.V. *The Apples of England*, Crosby Lockwood, London. 1946. 3rd rev. ed.

VARIOUS AUTHORS. *Fruit: Present and Future*, Royal Horticultural Society/Geoffrey Bles, London. 1966.

ANNUAL REPORTS OF EAST MALLING RESEARCH STATION (various).

ANNUAL REPORTS OF NATIONAL FRUIT TRIALS, Brogdale Experimental Horticulture Station (various).

PUBLICATIONS OF THE MUSEUM OF CIDER, Hereford.

Acknowledgements

COLOUR

The painting on page 119 is reproduced by permission of the Marquess of Bath, Longleat House, Warminster, Wiltshire.

Peter Blackburne-Maze 15, 38 top right, 62, 70, 71 bottom, 78 bottom, 87 top & bottom, 94 centre left; Bridgeman Art Library 11, 55, 74–75, 110; Bridgeman Art Library/Victoria and Albert Museum 22, 39; British Library, London 31; National Fruit Trials, Brogdale Experimental Horticulture Station 18, 38 top left, centre left, centre right, bottom left & bottom right, 71 top, 94 top left, top right, centre right, bottom left, bottom right, 95 top left, top right, centre left, centre right, bottom left & bottom right; E.T. Archive, London 35; East Malling Research Station 63 top & bottom; Mary Evans Picture Library, London 47, 83, 114, 123; Fine Art Photographic Library, London 6, 42, 90; Photographie Giraudon, Paris 27 bottom; Lindley Library/Eileen Tweedy 98, 103; Museum of Cider, Hereford 107, 122; Photoresources, Canterbury 26 top, 27 top; Photos Horticultural 79, 86; Scala, Florence 19; Brian and Sally Shuel, London 67; Spectrum Colour Library, London 78 top; Tate Gallery, London 58.

The painting on page 90 is 'Freshly picked apples' by Alphonse Mucha © DACS 1986

BLACK AND WHITE

BBC Hulton Picture Library/Bettman Archive 17; Janet and Colin Bord, Corwen 28; H.P. Bulmer, Hereford 104; East Malling Research Station 60; Mary Evans Picture Library, London 16, 21, 48, 82, 108; Fotomas Index, London 56; Lindley Library/Eileen Tweedy 25, 26 bottom, 33, 36, 44, 45, 50, 64, 65, 80; Mansell Collection, London, 8, 12, 13, 14, 20, 29; Museum of Cider, Hereford 100, 102, 105; Sothebys, London 109; Victoria and Albert Museum, London 120; The National Trust, Waddesdon Manor 124.

Index